PRAISE FOI
LEAP TO WHOLENESS

"With science, story, and vulnerable reflection, *Leap to Wholeness* provides a clear path to expanding into greatness through the magic of synchronicity and experiencing the curriculum of life with awe and wonder."

—REBECCA VILLARREAL, author of *The Amazing Adventures of Selma Calderón*

"The bad news is that *Leap to Wholeness* is not light reading. The good news is that if you can focus 100 percent attention on Sky Nelson-Isaacs's insights, even for a few minutes, you will reap rewards that will pay dividends for a lifetime. There is gold in every sentence and each time you read the book you gain new ideas that will help you become a more complete person who is in harmony with the world."

—DR. MICHAEL SOON LEE, coauthor of *Cross-Cultural Selling for Dummies*

"Based on solid scientific foundations from quantum physics, Nelson-Isaacs takes readers on a path of a holistic cosmos and inner healing. Like our outer world, our inner world also seems to follow the multidimensional entangled histories of a holographic universe."

—MARCIN NOWAKOWSKI, PhD, Gdansk University of Technology

"Nelson-Isaacs is a musician, poet, and physicist, and he gets them all mixed up so that his physics reads like poetry, and his poetry reads like music…. He gives us a rich understanding of the wholeness of our nurturing and human-friendly cosmic home."

—ALLAN COMBS, PhD, author of *Consciousness Explained Better* and *The Radiance of Being*

"... a great use of current scientific knowledge sprinkled with personal stories to help readers relate to that knowledge. This is a great contribution and a niche worth filling in the literature."

—WOLFGANG BAER, PhD, founder of Nascent Systems

"In this book, I feel humanity's yearning for a deeper encoded inner truth to guide us to a world that works interdependently, bringing out the best. Nelson-Isaacs doesn't demand change—he wisely sets the process clearly in front of us, nudging us along or catching us in a wave of wonder as we contemplate a world that really does work."

—REV. BARBARA LEGER, author, speaker, and founder of TEMENOS Center for Spiritual Living

"[Nelson-Isaacs] awakens our spirit by sharing his own insights and journey in a deep, authentic, and very personal way.... For those of us that have always wondered if there are coincidences, Sky's science-based blended with soulful approach answers this big question once and for all! Read on...there's so much more inside..."

—DANA LOOK-ARIMOTO, *Settle Smarter* podcast, executive coach, speaker, and author

"In this manual for more effective living, Nelson-Isaacs helps facilitate healing as he guides the reader to an awareness of not only their Wholeness but of the Wholeness of it All. A must-read for anyone looking for a deeper, more meaningful, and purposeful experience of life."

—REV. RAYMONT L. ANDERSON, DD, PhD, MSCD, Senior Minister of the Center for Spiritual Living Greater Baltimore

"Read *Leap to Wholeness* to experience a crack in your consciousness toward becoming more human."

—FRANCESCA MCCARTNEY, founder of Academy of Intuition Medicine and author of *Body of Health*

LEAP TO
WHOLENESS

LEAP TO
WHOLENESS

How *the* World Is Programmed *to* Help Us Grow, Heal, *and* Adapt

SKY NELSON-ISAACS

FOREWORD BY GEOFF THOMPSON

North Atlantic Books
Berkeley, California

Published by
North Atlantic Books
Berkeley, California

Cover art © gettyimages.com/mim.girl
Cover design by John Yates
Interior design by Happenstance Type-O-Rama

Printed in the United States of America

Leap to Wholeness: How the World Is Programmed to Help Us Grow, Heal, and Adapt is sponsored and published by the Society for the Study of Native Arts and Sciences (dba North Atlantic Books), an educational nonprofit based in Berkeley, California, that collaborates with partners to develop cross-cultural perspectives, nurture holistic views of art, science, the humanities, and healing, and seed personal and global transformation by publishing work on the relationship of body, spirit, and nature.

North Atlantic Books' publications are available through most bookstores. For further information, visit our website at www.northatlanticbooks.com or call 800-733-3000.

Figures 4.1, 4.2: Sky Nelson-Isaacs and Amalendu
Figures 5.1, 5.3, 5.4, 5.5, 5.6: Alexander Barnes, www.alextheactualizer.com
Figures 13.7–13.22: Sky Nelson-Isaacs and Alexander Barnes
Page 251: Quotation from *The Essential Rumi* used with permission of Coleman Barks
Appendix B Figure B.1: Alison Maines and Sky Nelson-Isaacs

Library of Congress Cataloging-in-Publication Data

Names: Nelson-Isaacs, Sky, author.
Title: Leap to wholeness : how the world is programmed to help us grow, heal, and adapt / Sky Nelson-Isaacs.
Description: Berkeley, California : North Atlantic Books, [2021] | Includes bibliographical references and index. | Summary: "How we can rethink our lives and reality to remove our filters and realize the wholeness that is inherent in ourselves and in our world"— Provided by publisher.
Identifiers: LCCN 2020031800 (print) | LCCN 2020031801 (ebook) | ISBN 9781623175689 (paperback) | ISBN 9781623175696 (ebook)
Subjects: LCSH: Holism.
Classification: LCC B818 .N45 2021 (print) | LCC B818 (ebook) | DDC 149—dc23
LC record available at https://lccn.loc.gov/2020031800
LC ebook record available at https://lccn.loc.gov/2020031801

1 2 3 4 5 6 7 8 9 KPC 26 25 24 23 22 21

I dedicate Leap to Wholeness to my nephew Yonim Schweig. Yonim saw the wholeness in himself and used it to bring out the wholeness in others. He reached across historical and cultural chasms to help people recognize the wholeness in their communities. His choices continue to ripple outward, unlimited by time and space.

CONTENTS

PART 3

FOREWORD

This is an exciting time to be alive.

Can you feel it?

Are you sitting there now, with *Leap to Wholeness* in your hand (and *I imagine* an appropriate beverage at your side), anticipating the arcana that this writing is about to reveal?

Be in no doubt, the words in this book, expertly recorded in ink, empirically earned in the world of men, and scientifically supported in the halls of academe, *will* create a shift in you. They did in me, and after sixty years of hard investigation on this spinning planet, that is no easy feat.

I have been around a few corners, as my Irish ancestors like to say—I have dueled with demons in dark places, I have drunk the hemlock of dangerous philosophy, and I have been frightened senseless by biblical revelation in moments of heightened consciousness: I am not quick to awe and I am never easily impressed.

This book did impress me, and I am in awe of the science revealed between its covers.

I am impressed because it confirmed what my hard-won experience had only supposed (that anything is possible). I am awed, because it validated what I had already quietly suspected: that our potential to grow is boundless, limited only by our own courage and our own imagination.

I have achieved much in my life, and many would be impressed by the list of accomplishments on my Wikipedia page, but reading the words in this book, I realized that (relatively speaking) I am still a neophyte: I have so much more to learn, and so much more to do that I don't want to sleep, for fear of missing a minute or an hour or a day to further enquiry.

I am not a science nerd; it has never really excited me. Until reading this book by Sky Nelson-Isaacs, I had restricted my spiritual education to challenging, exoteric experience in the manifest word and deep esoteric investigation into the exegesis of biblical tomes. What inspired me about *Leap to Wholeness* is that everything Sky talks about regarding science—the multiverse, holism, reductionism, singularity, the quantum—is equally present in the hidden works of Torah, Kabbala, The Zohar, the Dhammapada, The Bhagavad-Gita, Christian mysticism, Patanjali's Yoga Sutras, and The Holy Quran: Everett's "many worlds formulation," coined in 1957, shares a startling similitude to the mind-bending Yoga Vasishta, believed to have been written as early as the sixth century.

When bleeding-edge science and ancient religion concur with honest and brave life-rigour, it is impossible not to be filled with optimism.

For instance: the author talks about holism, and *purpose* being a holistic property of an organization, that attracts corporation: "a group of people coming together with a purpose." In fact, he says "corporation does not exist without purpose."

In the Judaic liturgy, this purpose is called Logos, literally "the word of God" or the principle of divine reason and creative order. When we connect to purpose, a universal corporation collects around us—it conspires in the most miraculous and unexpected ways to make even implausible dreams a living reality. This works equally well if you are an individual trying to change a personal habit or a conglomerate hoping to transform the world.

Purpose is religious, it is scientific, and in the world of form, it is a powerful modality: it is didactic.

And if our purpose is to selflessly serve others, its power is magnified exponentially.

I have quietly suspected this truth my whole life, but reading that purpose has a scientific precedent somehow makes it more real, more acceptable; it transforms the half-known into the certain and allows me to capitalize on what before was only conjecture.

What also struck me about *Leap to Wholeness,* and its courageous author, is that by offering validation to our *quiet suspicions*—the truths we hesitate to publicly announce or openly practice—it gives us permission to voice our grand intuitions in the world, and the impetus to act on them.

Every blessed individual who has been gifted with an incarnate life, whether they live for one hundred years or one hundred seconds, has a part to play in the unfolding of this stunning cosmos. Each has a lesson to learn and a potential to fulfill. Even if that purpose is not immediately evident, or clearly articulated by the circumstance of their living, every human occurrence is vitally important, and profoundly relevant.

I always believed too, that access to truth on the Great Earth was "shared equitably among all participants," even though, on the surface, this might not seem to be the case. How wonderful then, to "feel" this intuitively, and then receive vindication from a book that landed in my life serendipitously and was written by a man who specializes in scientific synchronicity.

My lasting impression—as the author reminds us again and again throughout this book—is that change, if it is to be effected at all, must begin with us.

We are the geometric point in our own life-construction.

The self is the only point of reference that is relative to our success in the world. Science, faith, and hard experience seem to concur on this one imperative: the universe is touch-sensitive to each of us, it favors those who strive, it does not favor those who fail to act.

It is no small folly trying to fix the world if we are not yet capable of wiping our own nose.

This is the kind of book that you read and immediately want to send to all your friends. If I were a corporate man, I would put my employees at the feet of this wonderful teacher.

He will tell them the truth: *that anything is possible.*

The proof is evident in the way he writes.

He will show them the truth: *nothing is beyond our grasp.*

The proof is evident in the way he lives.

May this timely book and its studied author act as a positive intercession in your life, may it bring you blessings (as it has me), encourage your curiosity, and deepen your enquiry.

Geoff Thompson
Coventry, England, 2020

ACKNOWLEDGMENTS

The feeling that comes from completing this book is gratitude. I am grateful for the opportunity to write a second book, grateful for the democratic society in which I live, and even for the fact that our planet Earth is in the "goldilocks zone," with the right amount of oxygen and oceans filled with liquid water.

If there is one thing I don't take for granted, it is the presence of my two favorite people, Eliana and Dana Nelson-Isaacs. The glacial delay of gratification that comes with writing and publishing a book would not be tolerable without the daily joy that these two people bring me. Thank you for your patience. Thanks also to my parents and family, I am lucky to be in your circle.

I am grateful to Geoff Thompson for his heartfelt contribution in the Foreword; Alexander Barnes, for his substantial creative contributions to the exciting graphics; Rizwan Virk, Celeste Fields, Ritu Kaushal, Jodi Reynosa, Trisha Smith, Linda Palmer; and all those who contributed stories to make the book better.

Much of the research in this book was supported by the Foundational Questions Institute (www.fqxi.org; grant no. FQXi-1801) via funding from the Federico and Elvia Faggin Foundation. Thanks also to Jurgen Theiss at Theiss Research. The author is particularly grateful to Ben Savitsky, Jason Elhaderi, Eliahu Cohen, Jeff Butler, Justin Kader, Dominik Safranek, Kai Chung, Daniel Sheehan, Federico Faggin, Anthony Aguirre, Jurgen Theiss, Richard Pham Vo, and Maarten Golterman for helpful comments and feedback.

Thank you to my editor Matt Hoover, to John Yates for the cover design, and to Alison Knowles, Janelle Ludowise, Bevin Donahue, Tim McKee, and everyone else at North Atlantic Books.

Thanks for enjoying this amazing ride with me!

PREFACE

How do you get something out of nothing?

This question is fundamental to our worldview in both science and religion. It led to the theory of the Big Bang, in which the whole of the physical universe, including space and time themselves, erupted into being.

The same view is present in the creation story of the Old Testament, where God created the universe out of the Void.

I wonder if this question is sticky in our minds because it reflects our basic fears. We go to work each day because, if we don't, we would likely run out of money and lose our home. We buy a house for ourselves because, if we don't, we may not have a safe location to sleep at night. We fight for our constitution because, if we don't, our rights may be taken away. Is your mental picture of the world that of an empty stage on which is placed the scenery of life? That's certainly what I've absorbed from my culture, and it seems like the essence of the Big Bang and the Biblical creation story.

But is this the only question we could ask? Although the answers to *how* the Big Bang occurred and *how* God created the world are quite different, in both traditions the *question itself* seems agreed upon: "How do we get something out of nothing?"

Granted, there are other cosmological theories in physics that do not involve creating something out of nothing. For instance, the "steady-state" theory—in which the universe expands in size but matter is continually created in the empty space, leading to an eternal universe—was considered before the Big Bang theory gained popularity. The Big Bang is a theory based on significant evidence. It is not my interest to question the theory. It is my interest to question the *question*.

As a young physics student at the University of California, Berkeley, I was consumed by a different story. I had the chance to create a hologram in a lab and smile with glee as my lab partner and I witnessed an ethereal three-dimensional image of a rubber duck appear in ghostly red laser light on a photographic plate. In my spare time I learned to write programs in C++ in order to play around with Fourier transforms and fractals. These are mathematical techniques for finding or generating patterns such as trees, clouds, or music, and they are the mathematics behind holograms. I had also read *The Holographic Universe* by Michael Talbot and other books like it. Looking back on all of these, I detect a common theme: wholeness.

When we form a mental image of our surroundings as if it is an empty stage onto which players are inserted, we are thinking as reductionists. In other words, big things are made from little things. The Big Bang answers the question "Where did all the little things come from?" But in my explorations in college I was seeing examples of things that existed "as-a-whole."*

Cutting a hologram in half is like closing window blinds halfway. You can still see what is behind the blinds if you shift your perspective to the side and look at a different angle. Similarly, if you shift your view on a hologram, it moves across the page like the view through the window. The image exists as-a-whole encoded into the film.

A rainbow seems to demonstrate this wholeness as well. A rainbow is not a thing "out there" in the air. A rainbow is an optical image that captures the geometric relationship between you, the rain, and the Sun. Nobody but you will see the same rainbow, and as you move, so will the rainbow. A rainbow is not holographic, but it demonstrates some mysterious properties of light that caught my attention at an early age.

---

* By the phrase "as-a-whole," I am emphasizing that there is something essential about it that can only be understood when taking the whole object into consideration—the hyphens help make this point!

As a student of physics, I also learned quantum mechanics. I like to say that quantum mechanics is the study of what the world is doing when you are not watching. Consider an electron. In quantum mechanics, instead of thinking of an electron as a real object in the world, we describe it with a *wavefunction*. The wavefunction can tell you every possible property about the electron. What's special about it is that it can hold many mutually exclusive possibilities at once.

This doesn't mean the electron can be in two places at once. It means that both possibilities can coexist up until the moment you want a real answer. What happens to the possibilities that *aren't* chosen? We can't say. It's a poorly defined question in physics, sort of like "What was there before time?" It is known as *collapse*, in the same sense that picking the winner of a raffle is a collapse of all the possibilities. All the tickets are thrown away except one.

So what? If quantum mechanics is the study of what the world is doing when we are not watching, real life is what happens when we do watch. Real life is a specific result that emerges out of a collection of possibilities. Real life is a single branch of decisions carved out of a vast decision tree of possibilities.

When considering this aspect of quantum mechanics, as well as the examples of holograms, fractals, and light that so intrigued me in school, I began to wonder if there might be another way to ask the question that led to a different but equally valid answer?

"How do we get something from *everything*?"

This was the question that came to me. Collapse of the wavefunction is a description of how we start with an entire tree of possibilities but only measure one actual result. A hologram holds visual information about an image as it could be seen from every possible angle, yet when we look at a *specific* angle we get a *specific* view. When we look at clouds, we see white because *all* the colors are mixed together, but when a rainbow is there we can see the colors separated from each other. If I look at a part of the rainbow that is red, it is red precisely because yellow, green, and blue are *missing* from that spot of sky. By default I should see white—every color—but the beauty comes from the limitations. All of these things are interesting because of what has been taken away from the wholeness, not because of what has been added to nothingness.

If the assumption that the world came out of nothing reflected a basic human fear of death and starving, what worldview would this other question lead to? If I see my education and career as "selecting between possibilities" rather than trying to "fill a bucket," would this change my mindset? What about my relationships? At the key tipping point of an important conversation, would it help to realize that there are many branches representing the outcomes of the conversation? Then my job is more like "finding my way" to a good outcome, rather than struggling to say the right thing at the right time.

The shift to the question "How do we get something from everything?" is a shift from control-mindset to flow-mindset. Our emotions are powerful influences on our behavior and often lead us into arguments and conflict when we can't manage them. If I ask myself "How do I get something from everything?" within a tense interaction, I find myself navigating it differently. I find myself focused on adapting to the flow of the situation, chiseling away what is unnecessary to find the solution that already exists.

Ultimately, seeing the cosmos as fundamentally whole seems to me a fundamental shift in human awareness. It is a step into personal wholeness as we discover how the filters we use obscure our hidden qualities. How did I become who I am out of all the possible people I could have been? How do my interpretations and filters limit the accuracy of my perceptions?

It is also a step away from isolation and into wholeness as a community. How do my inaccurate perceptions limit the potential for intimacy, vulnerability, and authenticity? As I've begun to learn about the filters I see the world through, I've gained more influence over how my life goes, even while relaxing control.

The most powerful tool that emerges from this worldview shift is the filter. A *filter* is anything that reduces something that is whole to a subset of its wholeness. Filters arise in optics in the study of light and data processing, they arise in psychology with the creation of the false self,[1] and they arise in quantum mechanics in the process of measurement.

Filters are familiar to students of Buddhism and other Eastern traditions of thought. In Buddhism the five aggregates of matter, sensations, perceptions, mental formations, and consciousness create impressions on us, impressions that color the raw reality and either create suffering or not. In Hinduism *Maya* is a "veil" over reality that fools us. Swami Satchidananda says everything in the world is part of Maya, and "your (manner of) approach makes it Sadgamaya or Asadgamaya (truth or untruth)."[2]

If we are seeking a world that works for everyone, our filters present an obstacle to achieving it. Yet we are each naturally whole, so finding wholeness—whether wholeness in ourselves, wholeness in our family, or wholeness in our society—is quite natural.

In my first book, *Living in Flow*, I examined how synchronicity leads to the experience of flow, or optimal experience. Here, we find that synchronicity and flow are means to healing, growth, and change. We seem to live in a world—a virtual-like reality—that is holographic in nature, reflecting that wholeness which we seek.

Alternating awe, wonder, and vulnerability, we will explore how our choices lead us toward or away from wholeness.

PART 1

THE NEED FOR WHOLENESS

Falling Down and Growing Up

The week before my college graduation I dyed my hair purple. Well, the color was called *eggplant*. I was set to receive my diploma in physics from the University of California, Berkeley, and after years of dedicating myself to my studies, eggplant hair at graduation seemed like a good way to rebel.

I also rebelled by not applying to graduate schools. I didn't want to continue to a PhD program in physics because my dream was to be a musician. Composing, recording, and performing were favorite activities of mine, which I had gained expertise in throughout high school and college. But rather than doing either activity—going to graduate school or actively creating the life of a professional musician—I slipped into a sort of "default path." I was hired as a graduate student instructor on campus and had a low enough rent to stay in Berkeley for another couple years, skating by.

I hoped that my natural talents would lift me up into a version of life I had fantasized about, but instead I found myself languishing. Rather than earning an advanced degree, or becoming a professional artist, or something else productive, this time after college became a time of struggle. I couldn't choose. I rebelled against the industry of physics, which I was afraid would strip away my creativity and preclude my ability to be the musician I was at heart. But I also rebelled against the music industry, unable to humble myself and create an identity that others could understand.

My struggle to find a place felt like somebody else's fault. But what I was unable to see at the time was how fear was guiding my choices. I felt that I was acting out of principle, but I couldn't see that my emotions were really pulling the strings. I wanted to gain traction but was scared to be seen. I wanted to make a difference but was afraid to be criticized. My inability to fit in was not simply a fierce commitment to independence and self-expression, it was a lonely retreat into isolation.

A few years later, as Bill Clinton was being impeached and everyone was starting to worry whether the world would end as the millennium flipped on Y2K, I had achieved no progress as a professional musician. I desperately wanted to break out of my self-imposed shell. I felt I had to prove myself by pushing the boundaries so far that somebody would notice. Instead of having purple hair, I shaved my head. I bought a van and fitted it with cabinetry and bedding so that I could have a home on the road. Then I packed up my instruments and set out for Los Angeles. The next six months were to be the most challenging of my life.

I didn't have any specific plans in Los Angeles. I sought out open mics to play music at. I scoured lists of agents and record labels and delivered press kits to their front desks. I gave a few performances and through lucky coincidences made some useful contacts. But within a couple months I was tired of hustling in my van. Dana, my best friend from high school, with whom I had shared a romantic, synchronistic rendezvous at the Louvre museum in Paris some years earlier, was also living in the Los Angeles area, and we had been talking together frequently. Our romance was reignited, and I soon moved in with her.

This was not the end of the struggle, though. My self-created difficulties were just beginning. Underneath my striving for success was an attempt at control. I felt lost in the world. My efforts to make a difference had failed over and over. I could see problems "out there" in the world that I wanted to help fix. I believed that by doing "inner activism" and becoming more self-aware, we could address environmental justice and social justice. Yet I kept tripping over my own problems, not making an impact anywhere. I wasn't in a place where I could help anybody else.

What were the patterns of behavior I was tripping over? At the time, I couldn't quite say. I focused my attention inward to try to overcome my own demons. I worked at it. I studied spiritual texts. I meditated. I fasted. I had faith that personal discipline would help me overcome my obstacles.

It was hard to see at the time that I was stuck inside a set of filters in my own mind. I thought I wanted to be recognized for my talents, but maybe I was afraid to fail. I wanted to take a stand, but maybe I was afraid that people wouldn't like me. I was set on making a difference, but maybe I would have been satisfied with feeling included.

Rather than talking positively to myself and building my confidence, I became more self-critical. I was sure that if only I achieved a greater level of discipline, I would break through to something greater. I pushed myself harder and spent time alone in deep thought without the benefit of a guide or teacher. I was not making enough money, but I tried to push this stress out of my mind. I kept fasting intermittently. I was convinced that if I could just break through, then everything would work out. But I found myself unable to sleep through the night, my mind too active to relax, and my body too undernourished to keep me anchored.

One morning in late winter I woke up before sunrise and drove to a nearby mountain, seeking inspiration in nature. I watched the sunrise and felt blissful yet untethered. Where was all this going? As I descended the mountain to my car, I became lost from the trail and ensnared in a grove of poison oak. I crawled through it on my hands and knees, and by the time I made it back to my car I was alarmed. I drove home to find Dana worried where I had been. I told her I needed to get checked out by the doctor, and she agreed. She had seen me struggling and was relieved for some outside support.

Once in the emergency room, the nurses noticed I was severely underweight. Through my inconsistent spiritual fasting I had starved myself to a wraith. My mortality struck me the moment the hospital staff had difficulty inserting an IV needle in my arm. I wasn't invincible. My choices had brought me to the edge of collapse, not to changing the world but to the hospital. If I wanted to be somewhere else, I had to make different choices. I had shaved my head, lived out of my van, and rejected every form of stability that society offered. In doing so, I had isolated myself and lost my footing.

I could only imagine how hard that time was for Dana, the person who had been closest to me during this ordeal. I called my father, who flew down the next day to come get me. We drove home together in my van, and I embarked on a new chapter.

Once home I checked myself into a residential treatment facility in order to regain some perspective on my life. I did indeed find healing in the ten days I stayed there. My roommate was a man who had persistent tinnitus, or ringing in his ears, at a deafening volume that caused him ceaseless pain. In seeing the true suffering that others were going through, I saw how self-involved my own suffering was. I had been stuck inside my own mental filters through which I saw myself as "not enough," and because of that interpretation I had acted in self-defeating ways.

One night while staying there I dreamt I was in my childhood home. My best friend from childhood was with me, with a loving and kind presence, but I was banging my head against a wall. I woke up with a visceral sense of self-compassion. If I was to make more of my life, I was going to have to be kinder to myself.

During my stay in the facility, I learned new ways to manage my thoughts and emotions and avoid negative self-talk. I also made some new friends. I checked myself out of the facility and started the difficult work of rebuilding my life.

I had been seeking fame, but eventually what I found was deeper connection: with Dana, with my family, and with the other important people in my life. I took this seriously. I began to see the potential for greater freedom in my life, not driven by my childhood fears but instead having real choice. I had an opportunity to reevaluate which goals were important to me and which thoughts were motivating me. This was my first conscious experience of healing, the peeling back of patterns that had warped my perceptions and influenced my daily experiences for many years.

The Grief of Isolation

In the climax of the movie *Contact*, based on the book by Carl Sagan, the lead character is being shown the wonders of the cosmos by an alien who says, "You're capable of such beautiful dreams and such horrible nightmares. You feel so lost, so cut off, so alone, only you're not. See, in all our searching, the only thing we've found that makes the emptiness bearable is each other."[1] Maybe, I came to realize, I was living out patterns of hurt from childhood. Maybe my struggles for professional success were masking a plea for intimacy.

In my travels to Los Angeles, I felt I had to prove myself. In setting out to make a mark on others, I had made myself as alone as possible. In my youth, I had learned to protect myself against intimacy. When I had experienced disappointment and abandonment in my family and friendships, I developed a belief—a filter—that by working harder in isolation I might heal those wounds. Instead, this simply brought me more isolation. When I found myself frustrated and lonely in Los Angeles, what did I do? I strengthened my connection with Dana. But even then, I interpreted this as a weakness on my part. I deeply enjoyed our relationship, but the filters in my mind couldn't shake the feeling of failing on my quest to "make it." I was trying to be a man and accomplish my vision.

My filters kept me in isolation, but intimacy binds us into a whole that is bigger than ourselves. To feel truly connected involves feeling vulnerable. Yet vulnerability is difficult. Moments of vulnerability don't *feel* successful. In striving to be outwardly successful, I was not letting people into my humanness. I didn't really let anybody into my struggles. I didn't build a meaningful support network. Vulnerable moments are not moments of self-protection, self-aggrandizement, or fantasy. They are moments when we hurt visibly, when we face reality and our pain is on the surface. Our suffering serves as glue that binds our hearts together. That's when we feel connected.

That experience was hard on Dana too. She had been there with me through the hardest experience I've had. Our relationship has lasted because it is built upon openness and vulnerability. Throughout all my darkest, most isolated moments, she is someone I have remained vulnerable with. Knowing that we will be by each other's side through even the hardest circumstances has given me a feeling of security that has helped me mature as an adult and trust in ways that I couldn't as a kid.

Since that experience over twenty years ago, I have had a career as a high school teacher, and as a software developer, and obtained a master's degree. I got married and had a beautiful child. Yet I am still on a path I embarked upon those many years ago. I am still seeking to feel more connected. I still come up against fear and anxiety, and I still frequently get defensive and turn down opportunities for intimacy. I still get tangled in mental filters and interpret other people incorrectly.

I have also gradually increased the times when I see things more clearly, where I try out new interpretations and experience a greater sense of wholeness. For me, wholeness means both a clear sense of who I am and a feeling of connection to the people in my life and to the culture I live in. It means placing a stake in the ground to mark my territory, but not being in rebellion to everything. To rebel is about dividing the parts, but to find greater wholeness is to heal the parts that feel separate, isolated, and alone.

This type of wholeness extends beyond philosophy and psychology. There is also a wholeness that emerges in the study of the physical world through physics. Light, that ubiquitous paintbrush that colors our eyes and carries our communications across the planet, embodies the wholeness I am talking about. This is seen most vividly in the study of holograms, which is part of a field of study called *diffraction*. A *hologram* is like a three-dimensional, realistic-looking photograph. As you tilt your perspective on it, the image you see moves like a real object. In a hologram, there is a sense in which the whole has an identity that is greater than the sum of its parts. This wholeness also appears in the vibrations of sound recorded in your favorite Pink or Prince song, as these signals are transformed into the frequencies that define their distinctive voices.

Through numerous familiar examples—light, music, rainbows—we will see that the universe displays a fundamental wholeness that is not embraced by our current worldview. Physics describes an indivisible world, and I wonder if the wholeness we see outside reflects the wholeness we seek inside. Through this wholeness emerges a picture of the world as a maze of choices, a "holographic multiverse." We navigate the maze of choices as best we can, and in turn it helps us along the way by presenting situations that help us heal and grow into a greater sense of ourselves.

We call these experiences *synchronicities*. A synchronicity was defined by Carl Jung as a meaningful coincidence, or more formally, "a falling together in time … of events which are related to each other … meaningfully, without there being any possibility of proving that this relation is a causal one."[2] I will use a related definition that captures the importance of space and time in the picture: *A synchronicity is a chance event that feels meaningful because it leads to an experience you are seeking to have.*

Socks Just When I Needed Them

Story contributed by Kim-Char Meredith

I ran out the house in a rush on the way to church and realized I was experiencing an epic "sock fail!" I resigned myself to my clearly-too-small ankle socks crawling down into my shoes. I arrived at church and went to my regular seat—and there I found a package wrapped in cheery yellow tissue from my friend Kelly. Her lovely note ended with "now I can share a smile with you," and it made my heart happy. I opened the gift and found a pair of smiley socks!! Clean, dry, and not too small.

Seen in this light, the world appears strangely like a virtual reality video game. This is not to make light of it, for it is a serious game with real consequences. Indeed, it is a game of quests like those undertaken by knights in feudal times. I wonder if living from this worldview can provide a needed upgrade to our personal and collective operating system?

Playing the game is not about gaining success, money, fame, or power. It is also not about rejecting these things. Living our lives is like climbing through a tree of possibilities, endeavoring with each breath to make things work out well, avoiding pitfalls and rotten limbs, and sometimes reaching pieces of fruit. It is about learning to see the threads of meaning in life more clearly so that we can make the best of each situation instead of creating further difficulties for ourselves. In this view, life is a game that we can play but never win. Instead, we are each on a quest to level-up our skills. The holographic multiverse that I will describe provides a backdrop on which we attempt to resolve our misperceptions, and as we resolve them our life takes on richer hues. This is a process of healing our wounds and becoming more whole.

In playing the game, the hope is that we can catch glimpses of the threads of meaning that are woven within the defining experiences of our lives. Rather than becoming more cut off from the world (as I did in my twenties) we can become more whole by embracing all parts of ourselves and expressing ourselves authentically with others.

It can be hard to be vulnerable. Fear of exposure is compelling. Fear of physical pain makes me want to choose the safe life-path, and fear of emotional pain makes me censor my speech. Yet by sharing my pain I feel more connected than before. By being vulnerable with others about my insecurity, I can dissolve my feelings of being isolated.

Playing the game is not about figuring things out once and for all. It is about getting onto a path of maximum growth, maximum satisfaction, and maximum connectedness. It is about healing ourselves from the filters that hold us back so that we can live our lives fully. In our quest to understand the process of healing, growth, and adaptation, we'll explore both the personal aspects of choice and the science of it. In the next chapter, we introduce the central theme of this study, the inherent wholeness found in nature.

2

A HOLISTIC COSMOS

In the first chapter, I shared my own personal story of breakdown and healing and showed how universal that experience of brokenness is. Now we are ready to look into wholeness as a fundamental property of nature, preparing us for the introduction to the holographic multiverse.

This chapter and a handful of others in the book focus on concepts in science that may be very unfamiliar to you; it might take a little more effort to follow the thread. Don't get freaked out! I try to mix things up, so if you find yourself struggling to understand, just smile and skip ahead. You do not have to take in all the scientific details in these chapters to get value out of the chapters on healing, growing, and adapting.

For this reason, the book has been divided into three parts. Part 1 involves some foundational science that is useful but not crucial to the overall story for the nontechnical reader. Part 2 is more readily accessible and emphasizes what can be gained at a personal level from shifting one's worldview to the holistic one. Readers can jump straight from here to part 2 if they so desire. Part 3 returns to the technical aspects of the science of the holistic cosmos in detail; it will hopefully be quite enjoyable for those who enjoy the scientific details of the discussion. Let's dive in!

What Is Holism?

What does science have to say about connectedness? Healing means getting on a path toward becoming a fully expressed human being.* Expressing ourselves fully is about wholeness. To embrace a greater sense of wholeness, we can start by immersing ourselves in worldviews where wholeness is possible.

Science is a major part of my worldview. And even if one doesn't outwardly profess a love of science, we all rely on the benefits of science in the form of technology. Science, it must be admitted, has contributed to the impersonalization of the natural world by objectifying what it studies. But is science inherently isolating? Does the scientific method require us to separate the system from its environment?

Science as a method is often used to break things down into smaller pieces that can be studied. The word *science* is derived from the Latin *scindere*, meaning "to cut off." The method of cutting up the world in order to understand it and control it forms the bedrock of much of our technological success. This idea that big things are made from little things is known as *reductionism*.

Reductionism shows up in biology when we reduce living creatures to the cells that they're made of. It shows up in physics when we reduce complex laws into their simpler building blocks. And it shows up in psychology when we reduce human behavior to a compilation of a few basic urges.

This approach is so successful in technology and theoretical sciences that it's easy to accept it without question. From it stems the philosophy of materialism, in which everything is reduced to interacting particles and meaningfulness is denied. The field of chemistry based on the periodic table is an example of the tremendous success that reductionism can facilitate. Yet reductionism does not have exclusive reign in the

---

* My wife, Dana, introduced me to the idea of being a fully expressed human being, and I immediately understood it to capture what I wanted with my life. Rather than have a set of obligations to accomplish or earn or serve a cause, when we ask ourselves what it would mean to be fully expressed, we put ourselves on track with flow.

physical sciences. For instance, an electron's location around an atom is described by an "orbital." The orbital is a pattern that looks somewhat like a flower. But the electron doesn't exist in a *specific location* predicted by the orbital. Rather, the pattern exists as-a-whole. It cannot be cut in half, and so we can't reduce the electron down to a simple point in space.

I envision that we are at the beginning of a new era in science. New worldviews are emerging. From complexity and systems theory to design thinking to quantum mechanics, we are beginning to see how things work together in mutual dependency. This seems no accident as the internet has become the glue of society. Central to this new era in science is a new understanding of wholeness.

I will show that there is a familiar example of a system in physics for which reductionism isn't true—light—and this straightforward example points to an underlying interconnected reality. Light is the center of our lives and the most important field of study in physics. It has either been a key player or played a supporting role in nearly every major advance in physics.

How can we tell that light is not reductionist? The evidence has to do with the two equivalent ways of describing light. One of these descriptions you are very familiar with. The arrangement of pixels you see when you look at a photograph or a computer screen is the familiar depiction.

But there is another way to describe light, known as the *frequency domain*. It is the frequency domain that allows a hologram to store three-dimensional information about a visual scene, say of an eagle flying in the air. A regular photograph of an eagle, for instance, is like a map with a one-to-one relationship to the eagle. By contrast, the frequency domain does not have a one-to-one relationship with the eagle. As we shall see in chapter 13, the frequency domain stores patterns that exist across all of space. Because of this, you can cut a hologram in half without losing half the picture. A hologram essentially stores information as-a-whole. To put this another way, if you want to erase a single part of a holographic image, you couldn't do it. The information about each pixel that you would want to erase is stored everywhere throughout the hologram.

The belief that each thing is separate from every other thing is so ingrained in our thinking that we are forced to name counter-examples as if they were exceptions to the rule. Yet light is so fundamental that it seems reasonable to question whether the behavior of holograms is just an exception or whether it shakes up the reductionist foundation.

What is the opposite of reductionism? What can there be besides the sum of the parts? This is not a straightforward question. In seeking to answer it one might ask, "Is there something that emerges when certain combinations of things occur?" Scientists and systems thinkers Fritjof Capra and Pier Luigi Luisi explore this question deeply in their book *The Systems View of Life*. "Systems thinking," originally credited to Ross Harrison, has identified two unique factors that emerge out of systems independently of their parts: configuration and relationship.

Configuration refers to a specific arrangement of the parts that allows for certain relationships to occur. The arrangement precipitates the emergence of new properties that are not present among the individual parts. Yet reductionism still has a place, say Capra and Luisi: "Reductionism … is fine when it limits itself to structure and composition. Emergence assumes its real value at the level of properties … based upon the proposition that the emergent properties cannot be reduced to the properties of the parts. On the one hand, we are saying that biological life is chemistry only; on the other hand, we also state that the arising of life as a property cannot be reduced to the properties of the single chemical components."[1]

The authors argue for a description of life called *autopoiesis*, or "self-making," based on work by Humberto Maturana and Francisco Varela.[2] The structures of life themselves are not living. "Life … is an emergent property—a property that is not present in the parts and originates only when the parts are assembled together"[3] in a particular configuration, to create a particular set of relationships. "The product of an autopoietic system is its own self-organization."[4] It seems evident that the ability of a system to spontaneously organize itself into meaningful relationships implies the existence of something greater than the parts alone.

Central to this definition of life is the boundary created between the entity and its environment. Using the definition prescribed by Maturana and Varela, Capra and Luisi say, "The main characteristic of life is self-maintenance due to internal networking of a chemical system that continuously reproduces itself within a boundary of its own making."[5] The systems view, when applied to life, "implies looking at a living organism in the totality of its mutual interactions ... Life is not localized: life is a global property, arising from the collective interactions of the molecular species in the cell," according to Capra and Luisi.[6]

Living systems take in resources from outside their boundaries, and they emit waste resources back out through their boundaries. As the authors say, "The behavior of a living organism is ... (not) determined by outside forces, it is determined by the organism's own structure—a structure formed by a succession of autonomous structural changes."[7] Therefore, "the perturbations of the environment do not determine what happens in the living—rather, it is the structure of the living being that determines what occurs in it," Maturana and Varela say.[8]

A process cannot occur without the parts, yet the parts alone do not tell us anything about the process. Life is thus an example of a "holism." *Holism*, which can be thought of as top-down thinking, describes a system whose whole is greater than the sum of its parts, a system that has qualities that depend on its entire configuration.

Holism does not exist "instead of" reductionism. Both can be used to successfully describe their respective aspects of the cosmos. Even if the world has ways in which it is holistic, the reductionist aspects of chemistry still provide a useful perspective, albeit within limitations. Atoms still make up molecules, and people are still made of organs. If you add a proton to the nucleus of a nitrogen atom, you obtain an oxygen atom. One is crucial for the creation of proteins and DNA, while the other is essential for the burning of fuel in biological organisms. It is simple to see the difference by simply tracking the parts, as if atoms were sculptures made of Legos. If you remove the liver from a biological organism, drastic consequences result, and it is easy to see many of these consequences through the reductionist lens. For instance, without the liver, your body can no longer process alcohol in the blood. Either in changing the atom or removing the liver, one could create a complete list of causes and effects.

Yet are these complete pictures of these systems? The introduction of a new proton into the nucleus of nitrogen alters the entire wavefunction of the entire nucleus and the entire atom. The structure of the entire entity changes, so *there is a sense in which no part of the oxygen atom is exactly the same as it was in the nitrogen atom.* Or when the liver is removed, a complex cascade of effects occurs. The liver is a part of processes that involve the whole body, and if we ignore these relationships, even those that seem unimportant for practical purposes, we may ultimately be missing relevant information. Holism can present a description that reductionism misses, while the idea that "big things are made from little things" provides powerful predictions that holism can't make sense of. Both are necessary.

Another way to get a sense of the holism we have discussed so far is to consider a corporation. The word *corporation* comes from the Anglo-Latin word *corporacioun*, "persons united in a body for some purpose." Clearly a corporation couldn't exist without the people who work there. The people are like atoms that behave in particular ways to make the corporation work. In the reductionist analogy, the corporation is nothing but the sum total of all the people who work there and the rules that they are supposed to follow. But in the definition is also the mysterious phrase "for some purpose." Where does purpose fit into this reductionist viewpoint? Purpose is something that emerges from the synthesis of the parts. It is the collective sense of urgency that arises around a looming deadline. It is the creative flexibility of rules that can emerge in situations when a threat is at hand. It is the hesitance of whistleblowers to speak their mind when they discover something to be amiss. Purpose exists when the parts are together, and it does not exist when the parts are separated or out of relationship. *Purpose* is a holistic property of an organization.

Charles Eisenstein describes the reductionism inherent in modern thinking as a fundamental story that we believe about ourselves. "The essence of the Story of Separation is the separate self in a world of other. Since I am separate from you, your well-being need not affect mine."[9] He identifies this narrative as a fundamental source of modern dilemmas like climate change. "Perhaps what we are facing in the multiple crises converging upon us is a breakdown in our basic problem-solving strategy, which itself rests on the deeper narratives of the Story of Separation."[10] We will explore the extent to which a different story in science, that of wholeness, can present a more effective strategy.

The Holographic Paradigm

The Scientific Age began around the seventeenth century, with Galileo Galilei's systematic experimentation on rolling and falling objects. This first era of science, based upon what is known as "classical mechanics," we will call the Classical Mechanics Paradigm, or Classical Paradigm. It began with Galileo, Newton, and Copernicus studying dynamics (the principles of motion), and lasted through Ludwig Boltzmann, Lord Kelvin, and James Clerk Maxwell, who contributed to the development of thermodynamics (the study of heat and temperature) and electrodynamics (the study of electricity, magnetism, and light). In the Classical Paradigm, the world is understood to be made primarily of physical material, and the relationships between things are thought of like machines in a factory. Space and time are fixed stages upon which the objects of the material world play out their particular circumstances.

In the early twentieth century, Albert Einstein, Max Planck, Niels Bohr, Erwin Schrödinger, Werner Heisenberg, and many others developed what we call modern physics, consisting of general relativity and quantum mechanics. Although it is still not understood how general relativity and quantum mechanics relate to each other, it is generally expected that they will someday be understood together. We'll call this the Quantum/Relativistic Paradigm, or Quantum Paradigm for short. In this paradigm, the world is fundamentally unpredictable. Instead of being solid material, things in the world are described as potential with certain probabilities. Absolute truth is not so straightforward, and instead relationships and perspectives loom large in describing reality.

I suggest that a new paradigm is emerging out of the Quantum Paradigm. We have seen clues of this new paradigm as we've learned more about the fundamental role of information in the world. Information has become a fifth pillar of physics, in addition to matter, energy, space, and time. We have seen clues of this new paradigm in quantum entanglement, where two separate entities can share a link that seems to transcend space and time. We have seen clues of this new paradigm while looking deep into space and seeing the universe as it was in the distant past. This is predicted by our understanding of space and time according to the theory of relativity, but it may even point to a timeless and spaceless description of the world, as we shall see.* We have seen clues in Dennis Gabor's development of a new field of optics known as holography, where three-dimensional images can be captured on a two-dimensional piece of film. In a hologram, every part of the film carries information about the entire three-dimensional scene.

These clues all have in common a transcendence of separateness, an expansion of our concept of space and time. This is especially dramatic in holograms. Reminiscent of how two entangled particles can be separated a long distance and still share collective properties, a hologram can be cut in half and still show you views of the entire image. The image of a hologram is not fixed in space but moves depending on the viewer's perspective. Although this could be called the Information Age, the dramatic flair of the hologram captures the essence of this new paradigm, so I will call it the Holographic Mechanics Paradigm, or simply the Holographic Paradigm.

I will describe in detail the scientific path that leads to the Holographic Paradigm in chapter 14, "The Mechanics of the Holographic Multiverse." But here it is worthwhile to give an overview of the theory, so the reader need not wait to understand the main points of the book.

Programming a Virtual World

Even while being a work of fiction, the film *The Matrix* is useful to keep in mind when considering the implications of a holographic world. In the movie, the characters inhabit a virtual world that feels real. The essential characteristic of this virtual world, for our purpose, is that elements of the world do not have fixed histories. When the lead character Neo makes

---

*In particular, for light there is a sense in which there is no separation between the Sun and the Earth—they are in "virtual contact," separated by a "null interval."

a choice to go left or right into a hallway, the enemy Agents whom he is evading can appear in either location. It is as if there is a script that must be experienced—the Agents and Neo are destined to meet and do battle—but Neo has free will to choose his actions and therefore determine *how* the script unfolds.

Since the past choices of the Agents (walking down the left or right hallways) have already happened, you might assume they are *certain*. We'll call these *facts*. But in order to allow for Neo to have the freedom to choose left or right and still have a meaningful plot unfold, those facts about the Agents' whereabouts must be free to adjust in order to accommodate Neo's choice. Notice that Neo doesn't know where the Agents *were* prior to his choice, so when they appear in front of him he can't really claim that they *adjusted* to meet him. Rather than say that the Agents *changed* their past in order to accommodate Neo's choice, we say that the past choices of the Agents are *undetermined* from Neo's point of view until Neo makes his choice.

This description works fine if we stick to Neo's point of view. The world becomes like a first-person virtual gaming environment where the specifics of Neo's world are "rendered" for him, and the specifics of the Agents' worlds are rendered separately for each of them. But do they experience a consistent *shared* reality when they meet? It is challenging to imagine how each person in the scene can exist within their own virtual rendition of reality, have equal free will over their choices, and yet experience congruency.

This is similar to the challenges faced in multiplayer online video game design, as I discussed in *Living in Flow*. The gameplay is made consistent through the just-in-time sharing of relevant information, a technique called *optimistic synchronization*. When Neo encounters the Agents, only the information immediately relevant to their meeting is shared, enough to determine when and where they engage. Other information, such as which doors in the building might have been left unlocked and could serve as an escape route, remain in a state of suspended animation until that information is needed. Until such time, the virtual world evolves independently for each inhabitant.[†]

[†] This is known as *retroactive event determination*.

How is this possible? Time must play an important role, because the main question we are tackling is how to make someone else's past align with your present. The model I'll present here could be called the "holographic multiverse," and it involves a different way of thinking about time than the previous paradigms. In the Classical Paradigm, objects in the world, such as doors, chairs, and people, have an independent reality and fixed histories. Their past is set in stone, because they exist with or without you. In this paradigm a shared, objective clock is measuring the passing of time, and everything that happens can be referenced to its ticking. In the Quantum/Relativistic Paradigm the sense of time is personalized rather than objective, getting faster or slower depending on how fast you are going. In both of these paradigms, you can get away with accepting the sense that you exist in the present moment, even if your present moment is different from another person's. Time is a relentless ticking you are subject to. The present is all there is.

Thank Goodness for the Robocall!

Story contributed by Dana Nelson-Isaacs

I kept getting cell phone calls from an unknown source identifying themselves as our electrical utility company. They were threatening to disconnect our service if we didn't call back with our sensitive personal information. Suspecting it was a scam, I hung up and called the utility company directly. It was the same number, so I figured maybe it was legitimate. After a long wait I reached the receptionist, and he confirmed that there is indeed a scam in which the perpetrators mimic their phone number. I was about to get off the phone, happy to have avoided the scam, when it occurred to me to ask about our account while I had the person's attention. After he opened our file, it turned out that our electricity was scheduled to be shut off the following Monday! We had accidentally underpaid and missed the warning notices. Those scammers helped us out!

By contrast, in the Holographic Paradigm external objects are like a virtual reality scene that is *rendered* for your perspective as the viewer. Time is not only personal to your perspective, it is tied to your awareness itself. Awareness defines time. Only that which you immediately experience is actually happening. And yet all of history exists at once. We are children living on a wider stage than we have previously realized, humbled by the privilege of being alive right now, and connected to all those who came before us and those who will follow.

In the Classical Paradigm, you are confined to the rigid sliver of time called the "present moment." In the Quantum/Relativistic Paradigm, that sliver is expanded so that each of us is allowed a personal experience of time. In the Holographic Paradigm, we are not stuck in the moment but surfing a timeline from past to future.*

It All Depends on Light

We return now to physical science and focus on a subject I find endlessly fascinating: light. Holism exists here too, maybe the most fundamental example of all. In the living systems described by Capra and Luisi, holistic properties emerge from the configuration and relationship of preexisting parts. But in the case of light, holism is its very essence.

What is the nature of light? Experiments with light might involve a light bulb, a laser, or a star shining onto an object like a face, a lens, or a piece of film. Although theories on the nature of light abound throughout history—including cultures in Greece, India, China, Iraq, and Egypt—we will start with the theories of French scientist René Descartes and Dutch physicist Christiaan Huygens in the seventeenth century. They both contributed to the theory that light is a wave, which can explain the fact that it bends when it strikes another substance such as water. This is called *refraction*.

---

* Some aspects of the proposals I will put forward are novel and are unsupported or unknown by some mainstream researchers in the field. It is important that the proposals be tested—that they be falsifiable—for us to consider them seriously. An examination of possible experimental tests is provided in appendix A.

But there are problems with the theory of light as a wave. In the late seventeenth century, Isaac Newton found that he could better explain refraction if he broke light down into little pieces, or particles. Newton called them "corpuscles," and his theory was a radical departure from the wave theory of Decartes and Huygens.

But Thomas Young showed in 1803 that one can also think of light as-a-whole. He shone light through two tiny slits and showed that the light didn't just pass through one slit *or* the other, but one slit *and* the other. In other words, the light was spread across the slits, like the way a wave in a harbor is spread across the pillars of a boardwalk as it laps around them. It seemed that different aspects of light's behavior were better explained by different theories. For some situations the particle picture was better, but for others the wave imagery was essential.

One hundred years later, Swiss physicist Albert Einstein ushered in the new era of quantum mechanics by successfully explaining why certain metal surfaces can only absorb light above a certain color. He flip-flopped us back to the particle picture. A particle of light is called a *photon*. If you want to knock an apple off a branch, you could throw millions of ping-pong balls at it (a lower-energy color of light) and never knock down the apple. But just one well-aimed throw with a tennis ball (a higher-energy color of light) could do so. Einstein reasoned that lower frequencies of light were made of very unenergetic particles, and higher frequencies of light were more energetic.

Physicists had a lingering dilemma between these very different models of light. Through the development of quantum mechanics over the following decades, these pictures merged into a paradoxical and subtle understanding of nature called *complementarity*. Light, atoms, and everything else can be described in the language of waves—waves of possibility. These waves can tell you statistically where and when to expect the photon (i.e., light) or atom, but they don't tell you anything about where it *actually* is. But when you do measure the photon or atom, it seems like a particle. It will appear to have traveled in a single straight line—unlike a wave—and it will have a particular location—unlike a wave. So our modern view of light has become quite sophisticated, incorporating all the previous advances into a nuanced picture.

Imagine looking up at a star in the sky. The light you see coming from the stars has traveled many years before reaching you. Yet those photons are not traveling like bullets through space. Instead, think of them as ripples in a lake, spreading outward from the star in all directions. They are ripples of possibility, for the light doesn't know where it will be received. It sends feelers out—waves—in all directions, and then when it gets absorbed by your eye (or as a discrete pixel in a CCD camera on a telescope) its endpoint is decided and its path becomes clear. At this point your description suddenly changes, and light is more like a particle that has traveled in a straight line between the star and you.

Thus, before it is observed, light is like a wave with the potential to be in many places, but after it is observed light is like a particle with a definite location. This is called *wave-particle duality*, a central aspect of the field of quantum mechanics.

The Most Beautiful Example

Yet holism with light goes deeper than this example of wave-particle duality, and we knew about it long before the development of quantum mechanics. Take, for instance, the most beautiful example, a rainbow.[11] You perceive a rainbow far away in the sky. It always appears directly opposite the sun, when the weather is both rainy and sunny at the same time. And it always appears at an angle of forty-two degrees in the sky. The rainbow seems like an "object out there," with a left side, a right side, and a middle.

But what if I remind you that when you try to drive past the rainbow it moves along with you? You'd say, "Oh yes, that's true. A funny thing, too!" If you chase the end of the rainbow in order to find a pot of gold, you know that the end of the rainbow recedes away from you as you approach it. There is something really subtle going on with the light from a rainbow. It is clearly different from the trees or mountains moving past your window, for you can never catch up to a rainbow.

Throughout this book we shall explore the idea that a rainbow—and light in general—does not exist in the normal space you are used to but in *frequency space*. We may alternately call it *pattern space*, or *the frequency domain*. We'll use these terms interchangeably. It is spaceless.* When you move your body, the entire rainbow moves across the background with you, as if it were an object flying through space alongside you. Yet another person down the road sees the rainbow in a different place. When you look at a patch of sky and see red, the other person looking at the same patch of sky might see yellow, or violet, or just blue sky. In fact, just about any place in that region of the sky can appear to be part of the rainbow for somebody looking from the right direction.

It seems that the rainbow you see must not really be an "object," in the usual sense, but rather a *pattern*. A rainbow emphasizes the relation-ships between different parts of space, just as much about you as it is about the water in the sky and the position of the Sun shining the light. There is no physical thing sitting in the sky that is red, orange, yellow, green, blue, indigo, and violet. Rather, it is a collective pattern of light traveling from the Sun behind you to the raindrops floating in front of you, then bouncing to your eyes. Think of it as a *relationship*. A rainbow demonstrates a wholeness of the space in which you stand, not a thing but a *motif*, an expression of the *organization* of light.†

---

* *Spaceless* and *timeless* are difficult terms to define. First, realize that everything you've ever seen or done existed at some specific point in space and time. The distinctions of space and time provide a language for our very thoughts, locking them in place. Yet this is a crutch of sorts. Here we try to imagine the world through use of a different language.

† A similar, quite enlightening example was brought to my attention by my friend Justin Kader in private correspondence. "When walking along a waterfront as the sun is low over the watery horizon, the reflection of the sunlight takes the form of a streak of light that starts at the point on the horizon just beneath the sun, and ends where the water meets your feet. As you move along the water's edge, the streak of light along the water 'follows you,' and the streak you saw five steps back is not visible from your new vantage point. How could it be that the 'object'—that streak of light—could just disappear from the point where you originally saw it and re-appear at your new position five steps away? The answer is that the thing is not an object, otherwise it would be visible from a dif-ferent vantage point. There really are an infinite set of streaks along the water, available for any observer at the water's edge, simultaneously. These spectacular reflections are everywhere along the water's edge, but you can only see one at a time!"

To help you picture what I mean, imagine a rainbow in the sky above a still lake, as in Figure 2.1. You would see a reflection of the rainbow in the lake, would you not? No, you wouldn't. You would see a reflection, yes, but of a different rainbow. The rainbow you are seeing in the lake is a different rainbow than the one you see in the sky. The reflected rainbow is coming from a different region of the sky than the one you see directly.[12]

You don't even need a rainbow to witness the wholeness of light. Wholeness is how light operates everywhere in the world. You can experience it right now in your seat. Just take out a camera and point it at something nearby, say a bookshelf with variously colored books. You could capture just about all the details of the scene with one photograph. It includes some books that are blue, some that are black, some that are red, and so on. These colors are entering the camera at the point in space where the lens is. But if you move your camera just a couple inches to the right,

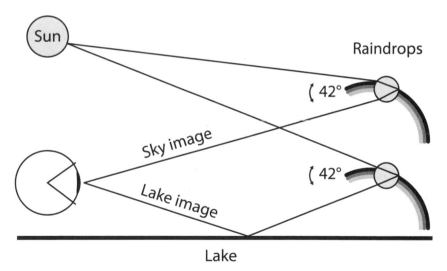

FIGURE 2.1. A rainbow is formed from raindrops reflecting sunlight of a specific color to our eyes. Each color of light reflects at a certain angle, averaging around forty-two degrees. If we look down toward a still lake while a rainbow is in the sky, we will see a reflected rainbow. However, we are seeing it in a different part of the sky, as is seen from the paths of sunlight shown in this figure. It is the reflection of a different rainbow!

you find basically the same colors of light hitting the lens at that point in space too. Light of every color from every book on your shelf exists at every point in the room!* The colors are not really separate; they form a complete mashup everywhere.

In fact, the image the camera sees doesn't even exist before it hits the camera's lens. Light "travels" from the bookshelf to the camera as a set of relationships in *pattern space*, and the lens converts it into an image you can see in regular space. Prior to hitting the lens, the information contained in the light is spaceless. Without a lens, it would be indecipherable. It doesn't exist in any specific location but rather is dispersed throughout the room.

This is truly an astounding—and nearly impossible to grasp—concept. We are so intimately familiar with space as our means of making sense of the world that imagining the world without it is like trying to think without knowing any words.

Yet we can bring space back into the picture. When the light enters the camera or your eye through its lens, the light becomes "spatialized." The lens orders the blue light from one book to go over here, and the red light from another book to go over there. When the light hits your retina, it is no longer diffuse, no longer spaceless. What you see on your retina is a one-to-one rendition of the objects around you.

Light shows us that all parts of space are woven together into a whole. In coming chapters we will explore physics without space or time, the most beautiful physics I know of. Before we get there, I want to explain how what we've discussed so far relates to the most mysterious thing I know of: synchronicity.

---

* Except, of course, where the line of sight is blocked.

3

SYNCHRONICITY AND THE HOLOGRAPHIC MULTIVERSE

Does the holistic nature of light have implications for our worldview and how we relate to the cosmos? In the coming pages we'll discuss the idea of the "universe" in a more general way—the "holographic multiverse"—and explore how experiences of synchronicity may be better understood within this model.

What Is a Synchronicity?

My friend Martha owned a ring with a blue stone in it. She had forgotten the name of the stone but liked the ring and wore it frequently. One day, her curiosity nagging at her, she took a photograph of the ring and emailed a jeweler she knew to see if they could help her identify it.

Later that day, she went to a nail salon to get her nails done. When the manicurist was finished, she asked Martha if she had a stamp card to earn credit for her visit. Martha thought she might, so she fished in her purse and pulled out a stamp-card-sized object. It turned out to be the business card from the jeweler who sold her the ring. Written by hand on the card were the words *blue topaz*. She had found the answer she was seeking.

Finding that card is an experience that wouldn't have been significant under normal circumstances. When the manicurist prompted her to look inside her purse, this was fortuitous because it led her to find the answer to an unrelated question that was on her mind. This is a synchronicity, a circumstance that felt meaningful to her because it led to an experience she was seeking to have.

A synchronicity is a chance event that feels meaningful because it leads to an experience you are seeking to have. In Martha's case, synchronicity helped her satisfy her curiosity. In other cases, the same phenomenon can help guide us in more profound ways and lead us to long-term healing and growth. I see synchronicity not only as a cute coincidence, like it was for Martha, but as a tool for personal and social change. We should not underestimate its potential impact on our lives and on our communities.

So what is a holographic multiverse and how does it relate to synchronicity? A *universe* is a collective term for everything that exists. A *multiverse*, then, is all the possible universes. It implies that what we experience is just one of many possible universes, branching like a tree into the unknown. A *holographic* multiverse, however, requires more explanation.

To answer this question, my research explores a field called the foundations of quantum mechanics, and in particular the vexing problem of time.[1,2,3,4] Physicists still struggle with an understanding of time in our theories, maybe because time is so personal to us. Our experience of it as "the present moment" defines our very lives and makes it hard to step outside of time and see it objectively.

It seems that in order to understand how meaningful coincidences—synchronicities—happen, one would have to allow time to be flexible. How did the right business card end up in Martha's purse? Or, alternately, how did Martha end up in a situation where she was asked to look in her purse? Rather than propose that she was reading minds or sending vibes—ideas not supported by what we know of physics—it seems more reasonable to me that our understanding of time is lacking. If the past is flexible, then maybe it could be adjusted in a meaningful way to match the circumstance.

But in physics, time and space are really one unified thing. I started thinking again about holograms and the amazing way they store three-dimensional images in space on a piece of film. If time is really the same as space, does that mean holograms could help us understand time in a new way?

A hologram is a special type of photograph. Let's picture an eagle as our example image, for you may have seen a hologram like this on a credit card. When you look at some credit cards, there is a small, silvery-looking square with a picture on it, such as an eagle. The eagle appears to be three-dimensional, floating underneath the surface of the credit card, and rotates its perspective slightly when you tilt the credit card. The hologram on a credit card is used as an anti-counterfeiting device, since holograms are much more difficult to forge than some other types of watermark.

How is a hologram of an eagle different from a photograph of an eagle? As you are familiar with, a photograph captures a fixed replica of a two-dimensional scene. Say there is a picture of a flying eagle in the newspaper. If you look at the newspaper from the side, you'll notice the image gets skewed, but the perspective of the eagle that you see in the newspaper doesn't change. By contrast, when you look at a hologram of an eagle from the side, you see *a side view of the eagle.*

A hologram looks surprisingly real. It is realistic because it creates an effect known as *parallax*, so that as you shift your perspective on the eagle, its front and back portions move relative to each other. Rather than being stuck on a page like a photograph, the hologram gives the effect of floating above or below the film it is captured on. Its realism is quite startling. Holograms are popular in science fiction—the Holodeck in *Star Trek*, or the image that R2-D2 projects of Princess Leia talking to Obi-Wan Kenobi in *Star Wars.*

Compared to a photograph, a hologram captures a lot more information. A photograph captures only the average amount of light falling on the film, and as you know, you can only view the scene from the perspective that the camera was at when the photo was taken. But a hologram captures not only the average amount of light in a region but also its relationship to its neighbors. These relationships encode the three-dimensional information.

A hologram appears so real that it is not unreasonable to consider that our actual world behaves like a very real-seeming hologram. This might be something similar to *The Matrix*, where the characters find themselves in an utterly convincing simulation of reality. Yet it is only there for their benefit, with no history of its own, rendered on demand for the characters to enter into.

A World Rendered on Demand

It seems that a holographic world would be like a virtual reality or a simulated reality. In *The Matrix* the artificial simulation is created by a race of robotic machines to pacify humanity. What makes this simulated world different from the way we normally think about our world? It responds to choices! When the protagonist, Neo, chooses between route A and route B to escape a building, the *histories* of objects in the virtual world adjust. Suddenly, instead of being in hallway A, the enemy will be found in hallway B. History must be flexible to allow Neo and the other characters to have free will and choice. The adjustment occurs so that the plot stays interesting.

In *The Matrix* the purpose of the illusory world is nefarious. The machines want to fool human beings into compliance so that they can use the human bodies as an electrical power source. This is convenient for Hollywood. There's a simple narrative and an easy enemy. A real version of that—what I call a *holographic multiverse*—would function similarly but for a very different purpose. For our world to work like this, objects in the virtual world must be rendered for each of us individually. How could such a fantastical reality exist?

The key feature in common between the fantastical Matrix and the actual multiverse is its ability to respond to our choices and bring us meaningful experiences. Just as Neo's choice between hallways leads to different histories for his enemies, our choices lead to different histories for the people and things we come into contact with. This "flexibility of history" is built into the structure of physics, illustrated by an idea

---

Don't Steal My Cereal!

Story contributed by Iris

I had an appointment with the doctor, so I went downtown by bike. We happened to be out of breakfast options, so on the way to the doctor I stopped at the grocery store for muesli. When I got to the doctor's office I debated bringing it inside with me, but leaving it in the bicycle basket seemed fine since I wouldn't be long, and who would steal muesli anyway? I was very wrong. When I returned after the appointment the muesli was stolen, along with my expensive sunglasses. Bummer! Just so it wasn't a total loss, I decided to go back to the store to replace the cereal. I walked into the store and chatted with the grocer, who upon hearing my story produced a pair of sunglasses from behind the check stand and asked if they were mine. They were! They hadn't been stolen but lost, and if the muesli hadn't been taken, the sunglasses would have remained lost!

---

called "Wheeler's delayed choice experiment," which will be discussed at greater length in chapter 14. While the phenomenon has become quite well accepted in mainstream quantum mechanics, it is very controversial to apply these ideas to the macroscopic world. Yet here we will take that bold step and examine the consequences.*

I call the flexibility of an object's history *retroactive event determination.†* According to this principle, your reality only includes those things you have observed. Everything else is unformed, undetermined,

---

* I believe I will be forgiven for this conjecture as long as I can clarify for you, the reader, what is accepted science and what is speculative.

† Retroactive event determination is essentially a phenomenon that is well established in the microscopic world ("Wheeler's delayed choice") but applied to macroscopic objects. This is, perhaps, the most controversial of the proposals in this book.

impossible to pin down. If everything in the past was already fixed into place, then it would be hard to imagine how Martha could end up so easily finding the card with "blue topaz" written on it. But if the past wasn't fixed, then there were various possibilities. Maybe the card could exist in two *possible* places—her purse or her glove box—and when she went to the salon, history "decided" upon the first possibility over the second. Or maybe the manicurist usually skipped mentioning the stamp cards, but on this day they had coincidentally heard a radio show about how much customers value stamp cards, and this had prompted their comment. If histories were flexible enough, they could allow for many more meaningful occurrences than we expect.

At first this may seem like a utopian daydream. If we just did things "right," we could manipulate this system to experience positive coincidences all the time. But as we look deeper, we discover the hallmark of meaningfulness in all of life's experiences. We start to recognize every event as synchronistic—even disruptions and misdirections—because we see ourselves within a larger context of growth and healing. If we hadn't lost our job, we would have never found that new job opportunity. If we hadn't had that experience of financial struggle, we never would have been able to successfully build our business. If we hadn't gotten sick, we would have never opened up to a new level of intimacy with our family members. To have reached the end of life and not found the job we really wanted, started the business we always knew we could, or had a deeper connection with those we love—this would be the real tragedy. In retrospect we can often see the gossamer threads of a web linking many of our experiences together into a larger meaningful tapestry. The disruptions are merely synchronicities leading us deeper into flow. They are meaningful because they lead to an experience we are seeking to have, maybe not consciously, but in the deepest recesses of our soul.

So synchronicity doesn't simply provide entertainment value. It weaves a tapestry. Through meaningful experiences we grow in our ability to handle change, and we rise to our next level of challenge. Our lives become inherently more satisfying as we resolve problems that have plagued us since we were young. We become better able to deal with the complexities of relationship and express ourselves with compassion and power. We correct misperceptions, picked up in our youth, through which we have sifted our experiences. We become more adaptable, able to experience satisfaction and enjoyment at greater levels, and able to let grief and loss change us for the better.

Experiences of synchronicity can teach us about fear. When we react to a situation out of fear, we experience more resentment, more anger, more jealousy. We end up feeling less secure and less confident, and the quality of our lives seems to diminish. But if we don't shut down when synchronicities expose our vulnerability, we are shown how to allow ourselves to love others and be loved by others, making us less fearful and less reactive.

If the world were like a video game or virtual world that could be rendered personally for you in the moment, then it would be a game with a purpose. I am wary that synchronicity has anything to do with pleasure or success. Both pleasure and success are relative things and couldn't possibly be related to science at a fundamental level. Rather, synchronicity seems to bring richness to my life. Its purpose must have something to do with meaning-making.

Connection across Cosmic Distances

But what clues do we have that this is an idea worth pursuing? In this section we get our hands dirty with the clues that led me to research the holographic multiverse. The main thread will not be lost, however, for those who gloss over this section in favor of the next chapter, where we return to the inward journey.

Have you ever thought about what the world is doing when you are not watching? One day while driving home from work, the freeway was jammed up, so I exited and took a long and windy road home through the countryside. I was pondering this question and asked myself, "What do I see right now?" In that moment I could see a rolling hillside covered in green grass under a bluish sunset. But what lay beyond it? Surely I would know in a minute or so, when my car crested the hill. But I realized that my story of what lay beyond the hill was a fabrication in my mind. I could imagine the next hill, but the only reason I ascribe any reality to my imagination is that the next hill always appears when I get to the top of this one. My assumption that the next hillside exists is always verified, but it is verified by a *future observation*. In the *present*, all I know is what I am perceiving right now.

In the car that day, it felt like a breakthrough idea. To create a convincing virtual reality, all the designer would have to do is present one scene at a time to the viewer. In his book *The Simulation Hypothesis*, author Rizwan Virk refers to his similar insight as a kid while playing the Atari racing game *Pole Position*. When you played the game, there were drawings of mountains in the distant background. They gave the illusion of a wide world beyond the racetrack. But was that world really there? In *Pole Position*, the answer is no. The visual scene is rendered for the purpose of racing the car. There is no virtual world beyond what is presented.

I wondered whether space and time could really be like that, just data that we read from the environment. I have been intrigued by the mysteries of data processing since I was a kid. Data processing is the study of patterns. For instance, you can transform the recording of a song into an alternate form, its frequencies, to find patterns in it. Then you can easily pick out a single note from the jumble of sound, like a needle in a haystack. It has always seemed like magic to me. The transformation is done with an important mathematical process called a *Fourier transform*, discovered by Joseph Fourier in the nineteenth century.

The same mystery applies to digital images. In Adobe Photoshop you can do the same things with colored pixels, laying bare the patterns present in the pixel layout of an image. For instance, you can easily adjust the graininess by making changes in the frequency domain.

This mathematics is the same as that used to create holograms, so a seed was planted in my mind that day. A real physical object such as a hillside of grass could be described as information in the frequency domain. There could be a "spectrum" that captured all the information about the position and movement of every blade of grass.

In graduate school some years later, the idea began to take root. As I studied quantum mechanics, the Fourier transform showed up again and again. Could the frequency domain be considered just as real and important as the regular physical world we are used to? Could there be a "space of frequencies" like there is a "space of positions"? In one of my textbooks the author wrote, "Apparently the mathematics we have developed (for quantum mechanics) somehow 'knows' Fourier's work on integral transforms."[5] Could the Fourier transform be more fundamental even than quantum mechanics?

When we Fourier-transform a music file, we rearrange the sounds in time and re-express them as frequencies or patterns. The same is true with the spatial layout of pixels in a photograph. We completely remove time and space from our description, in favor of patterns. Is it then possible to remove time and space from our fundamental description of physical reality?

There were more enticing clues. In another textbook the author describes the mathematics of the frequency domain as "space-invariant."[6] This seemed to me like another way of saying "spaceless," where our usual understanding of separations doesn't apply. It got me thinking about light traveling through space. Light leaving the Sun supposedly takes eight minutes to get to Earth. This is true from our perspective as human beings measuring the distances, but is there a way to describe this process "spacelessly"?

Indeed there is. In physics it is known as a *null interval*, which is a fancy way of saying that if one could take the perspective of something traveling the speed of light, time and space would be meaningless. I discovered a mainstream academic paper written in the 1930s by chemist Gilbert Lewis suggesting that the Sun and the Earth are in "virtual contact."[7, 8] In this space-invariant picture, there is no separation between them. This even applies to light traveling from Andromeda galaxy, over two million light years away. We are in virtual contact with everything we can see, from people to stars to galaxies. While this term is not widely used, the idea itself is part of mainstream views of physics.

These queries led me to attempt to reformulate quantum mechanics with emphasis on the Fourier transform. Because Fourier transforms are pivotal to quantum mechanics, we are essentially saying "Look, when the fundamental particles go about their business obeying the laws of quantum field theory, they are actually acting out the characteristics of a giant hologram." The laws governing particle physics seem to be identical to the mathematics of holograms.

What Is the Underlying Reality?

What we will find in the latter chapters of the book is a deeper understanding of holism, how it manifests in the world, and how the laws of physics as we know them seem to be exactly those of a virtual world. Whether you want to think of our world as "virtual reality" or not is not really the point. The point is that the world does not persist when you are not engaged with it. Each successive horizon and all the blades of grass on it are like a movie set creating the impression of a wide world beyond. It will always be there when you check on it, but when you don't check, who's to say?

Not only is the movie set put together "on the spot," the distances of separation in space are a fabrication as well. They are like the stories your favorite aunt or uncle tells you: the narrative takes a while to tell, but it exists all-at-once in the mind of the storyteller. We are in immediate virtual contact with the Andromeda galaxy, and yet when its light travels to us the story takes two million years to unfold.

The "mind of the storyteller" is *pattern space* (i.e., the frequency domain). We will find that there is not just one notion of space and time, but two. The space and time we are familiar with I call the *coordinates* of things. Returning to the example of a hologram of an eagle, the everyday experiences of life are like the image of the eagle you see in the hologram. The strange thing, though, is that the holographic eagle is an optical illusion. There is no eagle there. This makes me wonder whether the interactions we have in daily life are illusory in a similar way.

If the stage and props of daily life are illusions, what is really there? What is a hologram, really? When you stop shining laser light on a hologram, the illusion of the eagle goes away, and what remains is a collection of wavy *interference patterns.* (See Figure 5.2, p. 69, for an example of holographic film.) To picture these, imagine throwing a handful of pebbles into a lake and watching the ripples spread. These are the underlying reality of the hologram. If we want to describe the pattern of ripples on the surface of the water, we use the other description of space, what I call *parameters.* These ripple patterns don't look like the eagle at all, but they *encode* the eagle within them. The eagle appears in space because of an underlying pattern in the frequency domain.

Maybe, then, the underlying reality is like that of a hologram.

More sophisticated than a regular hologram, the holographic multiverse is a hologram in space *and in time.* The ripples of the pond capture all of reality, not only right now but over history. In the frequency domain, history exists in totality, all-at-once. We are momentary witnesses to just a slice of history, one branch on a tree of possible filmstrips, each filmstrip describing one way our life could unfold.

If the *universe* is everything that is really happening, the *multiverse* is everything that could happen or could have happened. The name refers to the **multi**ple uni**verses,** lined up like branches of a tree. In a hologram, we see just one view of the eagle, but all possible views are encoded there. Similarly, in the holographic multiverse we experience just one timeline— our own life history—but all possible timelines are encoded there.

Within this framework, the possibility of synchronicity can emerge from a proposal I call *meaningful history selection.* This was described in detail in my first book, *Living in Flow,* and is outlined in appendix B here. According to this theory, out of all the branches available to us, out of all the possible ways our lives can go, it is most likely that we experience things that reinforce what we already believe.

Though modeling the world as a hologram may seem like an abstract undertaking, it may have great relevance for our everyday experience. Too common today is the worldview that we must simply react to a world that is already unfolding. Life becomes about coping with change after it occurs, believing that "things are as they are."

Business leader Joseph Jaworski writes, "If individuals and organizations operate from the generative orientation, from possibility rather than resignation, we can create the future into which we are living, as opposed to merely reacting to it when we get there … Leadership is about creating, day by day, a domain in which we and those around us continually deepen our understanding of reality and are able to participate in shaping the future."[9] If we live in a holographic world, what we think of as the "real world out there" is not yet certain. The circumstances yet to unfold can do so in a wide variety of ways, and they may depend on our choices. The past is less constrained than we might think, waiting for us to make choices before the responses are determined. In a holographic world, there are latent possibilities hidden in every choice.

In this model, the metaphor of a school comes to mind, where each lesson is appropriate to the student's preparedness. This provides a sense as to why the flow state studied by Mihaly Csikszentmihalyi is so effective: we experience flow when we treat life like a curriculum designed just for us, at the perfect balance of challenge and skill. With this picture in our head, we will now look at the way personal patterns of thinking and behavior may impact our journey through the holographic multiverse.

4

WE EXPERIENCE
A FILTERED WORLD

Picking up the thread begun in chapter 1, I'd like to weave what we've learned about information, perception, and holism into our experience of personal choices. The central idea is that, like how laser light passing through a holographic film is filtered to form the illusion of an image, our own perceptual filters convey to us a specific version of reality that guides our choices.

Perceived Isolation

Loneliness, or "perceived social isolation," is increasingly acknowledged as an epidemic. A study in the journal *Heart* found that "poor social relationships were associated with a 29% increase in risk of incident CHD (coronary heart disease) and a 32% increase in risk of stroke."[1] Another study found that one-fifth of the population of the United Kingdom experiences acute loneliness.[2] The Association for Psychological Science published a study that found "increased likelihood of death was 26% for reported loneliness, 29% for social isolation, and 32% for living alone."[3] The heightened risk for mortality due to "a lack of social relationships" (whether reported loneliness, social isolation, or living alone) is greater than the risk due to obesity.

Why? Is there something we can do? Is there something we are missing?

There are certainly some moments when I feel more socially connected. These are moments when I have a conversation that accesses feelings I have been hiding. When some real feeling is triggered and somebody is there to share the experience, I feel connected. This correlation is clear. If I do not experience authentic emotion in a conversation, then I do not feel connected. If I do, then I do.

For me, combating isolation means discovering and sharing feelings that I don't usually share. My feelings remain private either because they are too sensitive or because I don't realize they are there. It was not until recently that I gained perspective on some of these behaviors picked up in middle school.

I felt so lonely back then. I wanted to be liked and have close friends. Having close friends had been easy when I was very young, but in middle school I felt like I had to really work at it. To feel like I was in my friend group, I had to constantly monitor the other boys. I was looking for whether they were going to tease me or criticize me in some way. I was on the lookout for situations that would cause me shame or embarrassment. I tried to impress my friends so I would be accepted, and I hid my struggle from everyone. With each decision, I fought to be loved. It was hard to take risks because I simply couldn't handle being exposed in front of my peers. Everything I felt inside was too tender to share.

I recall the day after my favorite cat was hit by a car and killed. At school I simply had to cry. Tears were streaming down my face during recess, but I wasn't with my "friends." I was with another boy, Josh, who felt safe to talk to. He was a good friend to me and helped me through that difficult day. Then the next day I went back to my regular friend group and tried to fit in again.

Now, as an adult, I can see this tendency to hide. Instead of acting on it like I used to, I can share about it. I can learn more about myself even as I deepen my relationships with others. I am less compelled to try to impress others to gain the connection I want. I still do it, but when I notice it, I get honest and share about it.

Being vulnerable doesn't have to revolve around pain. Laughing serves the same purpose. Or sitting together in awe of the sunset. Or yelling at the TV together while watching the Super Bowl. Being connected is a personal choice. If what we want is to be known in our wholeness, then it is necessary to identify for ourselves who we are as whole selves. I think this is the struggle against isolation. It is a struggle to allow others to know who we are, and in the process to know ourselves. It is our striving to see ourselves reflected in the world.

Maybe we don't see our reflections accurately. Maybe, though vaguely aware of the turmoil inside, we don't feel comfortable letting it be known. Maybe the unknown inside us is simply terrifying to us. Is it the possibility of failure that we run from? Do our inner struggles feel like a burden to others? Maybe our hidden feelings scare us because they have powerful things to say. Maybe they want to say, "No, I will not stand for this anymore" to a victimizer whom we depend on for security. Maybe they threaten to undermine our security by doing things that will alienate others and leave us isolated.

But so often we become isolated anyway. We may "belong"—we become a participating member of society—but still feel invisible. If we leave part of ourselves out of the relationship, aren't we bound to end up feeling lonely? We are whole people, and if any part of ourselves has no home in the world we will ultimately feel that we do not actually belong.

If loneliness is defined as "perceived" social isolation, how much does our sense of isolation depend on perception? Self-awareness, or becoming aware of the alignment between how we feel and how we act, is a critical skill for thriving in the twenty-first century. People are more open to meditation and personal growth than ever before; according to the Centers for Disease Control and Prevention (CDC), the practice of yoga has increased dramatically in popularity among US workers over the past twenty years, nearly doubling from 2002 to 2012.

Because of our accelerating connectivity, many people today have a greater understanding of and skill with moderating their inner state. More and more human resources departments are bringing mindfulness training and personal growth exercises into the workplace. According to a 2018 study by the CDC, meditation practice in the United States more than tripled between 2012 and 2017, from 4.1 to 14.2 percent of the general adult population meditating.[4] We are privileged to grow up in a generation where it is more clear than ever that underneath these counterproductive behaviors reside out-of-control emotions. I believe it is possible now to shift the expectation that violence is inevitable by developing the skills to manage our emotions but not act upon them without doing so intentionally. What an exciting possibility!

As we become more aware of the personal choices that lead to feelings of isolation, we also develop a deeper sense of personal empowerment. People are throwing off the mantle of inherited identities in unprecedented ways, and running for positions of leadership in record numbers. People want change, and they recognize that it comes from themselves. The idea that "we are the ones we've been waiting for" was written by Hopi Elder Thomas Banyacya, Sr., as part of a prophecy many years ago,[5] and it came back during the 2008 political season as presidential candidate Barack Obama's campaign slogan. Since that time it has built momentum into a tidal wave of action. People realize that their participation is required.

The risk in this process is that if we focus solely on changing external factors, we might try to solve challenges such as climate change with the same qualities that have led to the underlying dilemmas in the first place. If we cannot resolve the things that isolate us—our tendencies toward righteous anger, manipulation of others, and desire to tightly control the outcomes of our efforts—our solutions will merely create new problems. In his book *Climate: A New Story*, Charles Eisenstein coaxes us away from our attempts to *conquer* climate change, for this perpetuates the war-like mentality that has created the problem in the first place. Instead, by healing oneself and returning to the joy-filled wondrous world that one experienced in nature as a child, one reconnects with their true motivation for being environmentally conscious. Climate change is a wound to be healed, not a problem to be solved.

I feel great sadness at what we have already lost. I was blessed to grow up among the forest and the shore, with fresh wind blowing on beautiful meadows. I feel sad that many people did not experience this personalization of nature and do not realize how it is an expression of their own soul. When we experience the fractal patterns of noise from the ocean or leaves rustling in the breeze, our mind is cleared. Our nervous system needs the patterns of nature to help it settle down.

The indifference we feel toward nature and the perception of it only as a financial resource are acts of disrespect toward ourselves. We are so much bigger than our immediate need for food or shelter, yet we have structured our communities to be in constant desperate struggle for these basic things. Even when these needs are met, we still see nature, other people, and even ourselves as resources to be exploited. Although there are many things that have made me anxious over the years of my life, life has been a beautiful experience overall, something to be appreciated and grateful for. I have gotten what I needed from life. I don't want my daughter to have less than I had. I don't want future generations to look back and wish that if only their ancestors had done more and had seen their filters more clearly, they would have realized the beauty that they were discarding into the trash can every day.

Yet I know that it is very difficult to understand the internal filters through which we speak and act. For instance, if I, as a white person, do not understand systemic white privilege, my ideas about climate change will carry white privilege forward into the world I want to create. According to Tatiana Garavito and Nathan Thanki's commentary on the climate change strategy of environmental group Extinction Rebellion, it "overwhelmingly reflects the concerns, priorities, and ideas of middle-class white people in rich countries of the global north. By doing so, it ends up silencing the stories of our (indigenous and PoC) communities, who for hundreds of years, have been resisting the root causes of climate change."[6]

Understanding climate change requires telling the truth about the impact of our economic choices and policies. "But whose truth?" ask Garavito and Thanki. "The economic structures that dominate us were brought about by colonial projects whose sole purpose is the pursuit of domination and profit. For centuries, racism, sexism, and classism have been necessary for this system to be upheld, and have shaped the conditions we find ourselves in ... We understand climate violence not as a threat of a future apocalypse but as the wind that fans the flames of existing injustices."

As a person of privilege, without understanding my culture's history and the stories of marginalized people, my efforts are likely to repeat the oppressive structures of the past. "The history of conquest, genocide, and slavery is the foundation of our modern economic system—the very system responsible for the global disaster that is climate change," writes Bobby, a UK activist.[7] But this is not just about studying the histories of other people. It is about studying my own inner history. What are the lessons I took away from my youth? How did I learn to respond when I felt threatened? What do I believe about the value of my voice? In what situations do I find it difficult to listen to others? Where in my life am I overly certain?

When we wake up to a new realization, it is often the case that we find ourselves surrounded by people who have already had that realization. There is something we could not recognize before, and yet with our new perspective we find that we are surrounded by a new community whose eyes we can only now see through. This is my experience in learning about white privilege. As I come to understand how my own thinking reflects my privileged position in society, I also discover many people around me who have already discovered this. I am the newcomer to the party. This is true of climate change as well. "Well before you started caring about polar bears and recycling, colonized and postcolonial peoples were already fighting to reclaim and heal their connection with the earth and all its life forms that were so brutally violated by European colonialism and extractive industries ... The climate movement, in the UK and globally, will be decolonial or it will be nothing," Bobby writes.[8]

Dealing with external problems is not separate from the internal problems we face. By seeking to understand wholeness, we are invited to continually turn our attention back to the inner causes of polarization and unhappiness. When we cease to feel isolated, we cease to allow ourselves to be polarized. Through connection, we find more happiness, and through happiness it becomes more difficult to marginalize the perspectives of others. The more we feel included, the more we are drawn to include others.

Then what are the inner causes of polarization and unhappiness? I will frame our wounds as filters. These are ways of thinking and feeling that limit our understanding of our experiences and lead us to act in ways that are not actually in our best interest. Our filters are tools by which we twist reality to match our own hurt, helping us justify our conclusions and, ultimately, avoid intimacy.

We seek not to be right but to connect. We seek not to punish but to forgive. We seek not to harden but to soften. We seek not change but healing.

What Are Filters?

I'm in a dream. I am floating in outer space, in orbit around the large globe of Earth.

Not too far in front of me I can make out the form of a musician. The musician is playing a piece of exquisite music whose every passage divinely touches me. There is a gentle rising and falling of an orchestral sound, yet there is so much more than an orchestra contained in the music.

I look at the musician more closely. They are neither man nor woman. They seem to be playing a saxophone-shaped instrument, except it's not exclusively a saxophone, it could be drums, or it could even be a part of their own body. They are merged with the instrument. The musician, like the music, is sort of ... undefined. It's as if there are pure emotions present in the sound, nothing lacking in the music, and nothing held back in the musician. Every possible human emotion is expressed.

Then a transformation occurs. Walls appear around us, and I realize we are no longer in the pregnant void of space but back in my bedroom on Earth. I am still standing and watching the figure play, except it has definite form now. It is a woman playing a funky old drum kit. The music no longer sounds "divine," it sounds mundane and has flaws in it. She is not playing "every possible emotion" but rather a 4/4 rock beat with an imperfect sense of rhythm, like a teenager practicing in their garage.

It's as if she started with a vast set of layered possibilities, but many layers have been peeled away with just one remaining. It began with the timeless totality of who this person *could* be but ended with a *specific* human performer at a *particular* moment in their journey as a musician. I sense in the music the frailty of a single human individual trying to capture the essence of the infinity that is their soul.

This dream illustrates the concept of filters. *Filtering* as I will use it here means the peeling away of possibilities. In the dream, I start with a musician whose form is hard to pin down, and music that transcends genre. Are they a boy or girl? Are they even human? Am I listening to gentle classical music or rock music with a beat? Is it the soft sound of a violin or the more punctuated sound of a saxophone, or drums? We can call these *superimposed possibilities*, as if many different stereos are playing many different types of music all at once. Think of the process of filtering the sound like turning off all the stereos except one. Which music is left? It depends on which stereos you turn off.

Similarly we can remove all but one of the possibilities for who the musician is. Drummer or saxophonist? It's like showing many different videos at once—the superimposed possibilities—and then turning off all but one. Filtering refers to any process that removes some elements from a collection, keeping only those that meet a chosen criterion.

We learned in the last chapter that a hologram contains the possibility of viewing a subject from any angle. When we look from a single angle, we filter the light we see. It shows us only the information left once the filter is applied. We shall find filters to be a central theme in our study of the holographic multiverse or virtual world that we seem to live in, and also a central theme in our examination of human nature and the choices we make.

In physics the most common use of the word *filter* is the processing of information or "signals." When you listen to the radio, your radio receiver is soaking up an extremely complex signal of electrical vibrations from the space around you. If you are tuned to 90.5 megahertz, your stereo resonates with signals that vibrate 90,500,000 times every second. Many other vibrations are present in the space around you, but only the 90.5 megahertz vibrations contain the signal you want to hear. Your radio removes everything else and then converts the remaining signal into sound. It may be local news, jazz music, or hip-hop. All the other possibilities have been removed.

I'll use the word *filter* in a number of related but subtly different ways. We filter our perceptions every day, making sense of the world around us. This includes interpreting the words people say, our mental impressions of the experiences that happen to us, and our emotional reactions to them. From these interpreting filters we choose actions that affect our lives.

A lens is a type of physical filter. You have a lens in your eye, and there are lenses in your glasses and in your mobile devices. The "mental lens," or mind's eye, helps us interpret the world, and it is a primary factor that distinguishes us from each other. Your perception of the world is different from mine, and our lives may be dramatically different as a result.

In physics a lens is a device that alters the amplitude and phase of the signals passing through it. It transforms one image into another, rearranging or selecting pieces of the whole, converting a raw waveform into a discernible image. Without a lens in our eye, our retina would receive only a completely white, washed-out impression of the world. With a lens in our eye we can see the features of our environment individually.

Our mind's-eye lens distinguishes our old selves from our new selves. While our genetics don't evolve much over the course of a single lifetime, and our skillsets change only a bit once we've reached adulthood, our perception of the world can change dramatically at any age. Our genetics and our opportunities to build physical skills are largely a function of luck, but perspective is completely our decision. Maturation is simply about changing our lens.

But it may be easy to forget that a lens is a filter. It selects certain information and focuses on it. Through the lens of traditional psychology, we filter for our neuroses and shine a light on them. We may lose sight of all the other qualities within us that we are not focusing on. Positive psychology practices, such as the experience of flow, refocus our lens on the personal qualities that help us thrive. Both approaches are useful to make progress in psychology, just as all the emotions we feel about a situation are important parts of who we are. We don't need to strip away our filters. Rather, the practice of healing our filters helps us use them consciously when needed.

This sort of mental filtering has a powerful influence in guiding the direction of our lives.

To this day I recall one specific meeting with one of my supervisors. I made an appointment to talk with him about receiving a raise. The conversation went well, and he was open to the proposal if I could come back to him with data to back up my request. Then he did something that lodged the event in my memory. He gave me a piece of negative feedback about my interactions with other people at work. He said it casually, not in the context of receiving a raise, but just as an opportune moment to discuss something that had been on his mind. Yet my filters kicked into high gear. What I heard in his statement was that I was flawed, and because of my flaws I did not deserve a raise. I connected the two conversations.

When this type of situation occurs, I have a choice. In one interpretation, I take the feedback personally, feel defensive, and ask for less money. In another interpretation, I bookmark the feedback but move confidently forward with my plan. This is the power of filters. Regardless of who my manager is, ultimately the choice of how I respond is completely up to me. Either interpretation may lead to quite distinct outcomes. On one timeline I earn more money and on the other I don't. What I choose depends on my relationship to the emotions stirred up in me.

The choices we make in response to our interpretations of life can change daily. Think of this like the weather. Our reactions and interpretations can seem like uncontrollable internal storms. But we can develop greater skill at mastering these storms. Filtering one interpretation instead of another may lead to substantially different actions on our part. Choosing between compassionate words and angry words can determine whether our interaction with a friend is a soft, comforting drizzle or a harsh, destructive storm. Any weather can serve us in different ways at different times, but by understanding how our filters shift the weather we gain some choice in the direction our life heads.

There is another, more abstract sense in which our lives get filtered and our personalities get formed. Like the person in my dream, we gradually refine who we are over the long term through the choices we make. Think of this like the climate. Who we become in the long run is related to the choices we make each day, just as the climate is a long-term average of the weather.

We'll look at these two processes of choice—the weather and the climate—through the physics of filters and the mathematics of holograms.

Navigating the Weather

How we navigate our inner weather is the most important skill of our era. In past generations, it was generally accepted that destructive behavior was inevitable. We accepted that war is necessary and that sometimes people are abusive. Yet we are coming to realize now that these are outcomes from the inner storms raging inside of us.

These storms come down to filters. How we filter both the outer information and our inner emotions determines the direction our lives evolve. When I receive feedback like from my supervisor in the earlier story, I choose my interpretation. Is he choosing to say this now because he doesn't feel I deserve more money? Or do I take at face value his claim that it was just a convenient moment to talk? It is totally up to me which pieces of information I choose to keep and which I choose to discard. Not all the information coming from the world or from my thoughts is correct. I perceive through filters.

Based on my filtered perceptions, I choose my response. This too is a different form of filtering. If we live in a multiverse, there are a plethora of branches evolving from each moment. If my supervisor's words hit a tender spot and I choose to say something defensive, I am choosing one set of circumstances. This is a timeline or branch of the tree. If I choose instead to receive the feedback gracefully, a different set of circumstances or timeline will unfold. These two branches are exclusive of each other.

Finding My Feet Again

Story contributed by Jennifer Redelle Carey

I had worked at a major bookstore chain during a difficult period when the company was going out of business. I finally left the job, and shortly afterward my mother died. Within a few years I also lost my father, and I spent a number of years without working in order to process these experiences. Eventually I took a crummy job for a while, then one day I said aloud, "I am ready to work in a bookstore again." This time I got a temp job at a small college bookstore rather than a large corporate chain. It was a great environment to work in, and I gave it my all. Within a few months I had earned a permanent position and the opportunity for a manager training came up. I was accepted into the training program, and soon thereafter a management position opened in the store, which I applied for and was offered. Then my dream of meeting Mihaly Csikszentmihalyi, founder of flow research, came to fruition when the bookstore hosted a conference in collaboration with his research department. The many synchronicities throughout this process—including working again with one of my managers from the previous bookstore—have brought me a sense of wholeness that I had partially lost through the traumatic experiences of the last decade.

The key idea of quantum mechanics is that all the possible properties of an object are contained within a mathematical structure called a *wavefunction*. One way of thinking about the wavefunction is to imagine it describing many worlds, or a multiverse. It is like a tree that branches again and again with every choice we make (see Figure 4.1). Each branch contains a single version of the universe. When you consider a whole collection of branches at the same level in the tree, they constitute a multiverse—many superimposed possible realities. On one branch, one set of choices is made and one set of outcomes is experienced, and on another branch a different set is chosen and experienced.

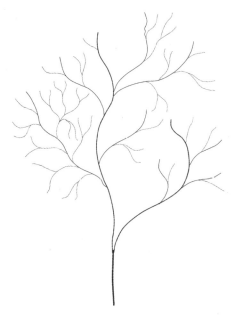

FIGURE 4.1. The term *multiverse* captures the idea that the experience you are having right now is just one of many possible experiences you could have. The different "universes" are spread out like branches of a tree.

The branching tree of our lives is like all the possible radio waves floating around you right now. You could turn on a radio wherever you are and tune in to—that is, filter for—any of the local broadcasts. Applying the same process to the tree of possibilities, we collapse the wavefunction from a many-branched tree to only one branch. This branch is the world we experience right now. Although the word *collapse* is typically used, I suggest instead that you think of it as trimming a tree. The possibilities you don't choose are carved away, and what is left is the world you experience.

Wavefunction collapse is still a mystery to physicists. Nearly a hundred years since its original development, we still do not understand why or how the multiverse goes from "all possibilities" to "the single possibility you experience." It's true that we can use quantum mechanics to successfully calculate the probable behavior of chemicals in a reaction or the magnetic properties of a subatomic particle. Yet our theories are always *probable*. We never know which universe will collapse from the multiverse; we only know a likelihood for each set of possibilities. How does the cosmos filter out all the other branches of the wavefunction tree, so that you experience only one of them? We don't know. We can predict for sure what the possible outcomes of an experiment are, but not which one will occur.

What is clear is that *how* we interact with the world impacts which choices will be available. In the microscopic world of physics experiments, we have some sense of how this works. In *Living in Flow* I presented a model called Meaningful History Selection; it theorizes that our anticipating certain experiences affect the probability of those experiences occurring.* Our daily choices define the possibilities available to us and influence their likelihood. Each branch of the tree is influenced by the weight of the apples on it, which appear on any branch that has the experience you are anticipating, as in Figure 4.2.

In the holographic multiverse model discussed here, how we filter and respond to our feelings selects not only a specific outcome but a whole extended branch. We experience not just a moment but a *timeline*. In *pattern space*, the world is timeless and spaceless, and as such there is no time or space separation between things. You are not simply choosing to solve a problem in the present moment. You are also timelessly connected to your future. All the branches of the tree are diverging filmstrips, and with each action you are setting yourself on a trajectory toward a target. Your choice of filmstrips includes not only the "here and now" but also the final frame.

This is the weather. We interpret our experiences and choose how we navigate the weather. In doing so, we trim off branches of the tree, filtering out a subset of possibilities for the future based on the choices of the past. How do our responses to the stormy or placid days affect the long-term arc of our lives? The climate we create through our choices determines the skin we live in.

---

* See appendix B for a short review of this theory.

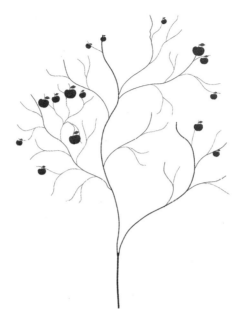

FIGURE 4.2. In the theory of meaningful history selection, pre-sented in *Living in Flow*, even when you miss an opportunity (an apple) there are many other apples further along the tree. When an opportu-nity is blocked, there are other branches of the tree to climb.

Influencing the Climate

Nearly a decade after my experience in Los Angeles, I was a full-time musician. I had left my job teaching physics to high school students in order to pursue my dream. But in 2008 I reached a decision point. Most of my earnings came from playing music for weddings, but my passion was writing and performing original music. I felt desperate in my desire for success.

I set up a gig at my hometown bar, hoping that it would be an easy win. It's an embarrassing experience to split a hundred dollars in tips with four of us in the band. I wanted this time to be different. But as the date approached, Hank, the event booker at the bar, proved to be unreachable. I felt the encroaching possibility of the event being a bust.

I felt angry and hurt that the people at the bar had so little invested in us. When the day of the gig arrived, we didn't even show up. I hadn't heard from the venue, so I assumed they didn't even remember it was happening.

The next day I received a phone call from Hank wondering what had happened. He was genuinely perplexed and disappointed. I had chosen a certain interpretation of the facts, and by not showing up for the gig, I also chose a corresponding filmstrip for my life. I had been wrong about how he felt, and my incorrect interpretation caused my inner weather to shift the course of my life.

This was the beginning of a period of four months that funneled me away from one dream and into another. I continued to feel frustrated and gradually stopped pursuing activities that would build my momentum as a musician. I thought to myself, "I shouldn't have to suffer like this. Let's see what happens if I just stop trying."

Where did I invest my time? I campaigned for Barack Obama, even meeting him face to face and handing him a music video I had created. At the same time, my father-in-law invited me to apply to work at his software company. While I intuitively knew that investing in this job reflected a shift in my dream, working as a software engineer had also been a dream of mine, and the stability of this path was appealing. I invested time preparing my resume and studying computer science.

Dana and I were also trying to get pregnant and were in the final stages of fertility treatment. These disparate threads came together in November. I interviewed with the software company the same week that Obama won the election, and a week later we found out that we were pregnant. Obama had been carried to victory partly by the oncoming Great Recession, and I was relieved when I was offered the job. December 1 was my first day.

I view this experience like a funnel. I had made a series of choices based on the weather of the moment—not showing up for the gig, not investing in the band, applying for the software job—that had culminated in a series of external events that shifted the climate of my life. Life moved more quickly than usual, with each choice building momentum toward one of the two possible outcomes. The apex of the funnel was like a tumultuous stretch of rapids on a river. I was jostled about by information coming at me—the potential pregnancy, the unfolding election, and the potential new job. In response I made the best choices I could in the moment, influenced by the weather systems in my mind, and was spit out of the rapids on a new path. Once I had emerged from the funnel, I had become a software programmer and a father, and we had a new president and a new economy. The weather had affected the climate.

It is natural to think of this process as the continual addition of layers. After all, by studying coding and creating a resume I added activities to my schedule, skills to my repertoire, and a document to my hard drive. But we can also think of it as a process of *chiseling away*. Out of the multiple branches of the tree, I landed on one by chiseling away the others. It could have been different. At each moment life is pregnant with possibilities. There is so much we *could* choose. If we take seriously the Holographic Paradigm worldview, then the act of choice requires chiseling away instead of adding to.

This brings us back to wholeness. We, as human beings, start out like the amorphous person in my dream. Many possibilities are present, limited certainly by our genes and other predispositions, yet still having access to a vast collection of potential personalities. We begin as wholes, and the way we interpret our experiences filters out the possibilities of who we become. We are not *formed* by the choices we make and the stances we take. We are *refined* by them.

What choices have you made that ended up becoming life-changing? Do you like who you have become through those choices? The process of changing the climate is like using your hands to dig a hole in wet sand for a sandcastle. You don't take a big shovel and scoop up the sand you need all at once. Rather, you gradually scrape away the edges of your hole, expanding the size of your castle's moat. With each pass, the hole gets a little wider, and with the sand you collect you can add to the top of the castle.

Our passage through life is a story of our choices and how they guide who we become. At every moment, untold branches of possibility spread above us. One by one, our choices remove possibilities from us, and our life direction shifts gradually. As we navigate the weather of our inner reactions and desires, these choices have a cumulative effect. Because of meaningful history selection, explored in *Living in Flow*, the situations that show up in our path tend to be ones that will help us grow, change, or heal in a way that reflects our choices. As we build momentum toward a certain target outcome, occasional opportunities will be decisive turning points on our journey. These are like gateways to another level of challenge, key moments when the climate changes and our personality takes shape dramatically.

As we travel along our path through the landscape of our time on Earth, the climate will change. We will grow and change in response to the challenges we are provided. Whether we evolve toward whole-hearted living or toward embittered frustration is a daily decision we make as we confront our inner weather. By dedicating ourselves to understanding our motivations and developing a growing mastery of our choices, we can shift the climate in a direction we will look back on and be happy about.

Chiseling Away from Wholeness

Distinguishing between "building from nothing" and "chiseling away from wholeness" is valuable because it urges us to switch from a mindset of lack to a mindset of abundance. If we are building from nothing, then we start with nothing. Everything must be *gained* through our personal effort, fighting against the odds. But if we are chiseling away from wholeness, then the emphasis is on *choice*. All the possibilities are there, and our job is to choose between them rather than tirelessly build from scratch. We must still take action, but our actions are not desperate and forceful but purposeful, thoughtful, and calm. The quality of our experience is different.

This distinction between emptiness and wholeness is present even at the foundation of cosmology. The current worldview is captured in the Big Bang, the most popular scientific theory describing the beginning of our universe. In the Big Bang theory, space, time, and everything in it spontaneously erupted into being out of the void. This is a well-established result with significant experimental evidence. But before we analyze its merits, let's back up again as we did in the preface and examine what question we were asking in the first place.

The Big Bang is the answer to the question "How do we get something from nothing?" Yet is that the best question to ask? Why do we assume it all started with nothing? The mathematics of filters invites a different question. When listening to the radio, we begin with a vast totality of radio information and filter it down to one station. Examining our world as part of a multiverse, we begin with all possible realities and whittle it down to just one.

Let's therefore frame a different question: "How do we get something from *everything*?" In other words, why do we experience only one specific set of circumstances from the totality of possibilities that could occur? To be clear, I am not debating the Big Bang theory. It's a powerful theory that fits the question it was intended to answer. Instead, I am questioning the relevance of the question that it answers. It's not that the Big Bang theory is wrong, but that it is correct only within a certain worldview—in the Holographic Paradigm, it may not be the only available description.

In *Synchronicity: An Acausal Connecting Principle*, Carl Jung describes how this filtering process affects the scientific endeavor at the most fundamental level.

> Experiment ... consists in asking a definite question which excludes as far as possible anything disturbing and irrelevant. It makes conditions, imposes them on Nature, and in this way forces her to give an answer to a question devised by man. She is prevented from answering out of the fullness of her possibilities since these possibilities are restricted as far as practicable. For this purpose there is created in the laboratory a situation which is artificially restricted to the question and which compels nature to give an unequivocal answer. The workings of nature in her unrestricted wholeness are completely excluded. If we want to know what these workings are, we need a method of inquiry which imposes the fewest possible conditions, or if possible no conditions at all, and then leaves nature to answer out of her fullness.[9]

The way we frame a scientific question filters the available options right out of the gate. This makes me wonder if our *assumption* that the world arises out of a void may limit our ability to understand anything about wholeness.

Wholeness applies both to the physical world and to human beings living in it. It implies that we ourselves start from *everything* and become *something*. Like Michelangelo finding David within the stone as he chiseled, through our choices we expose the edges of the sculpture we are becoming. If you look back on your life, you will probably notice that you have become who you are due to specific experiences that you've had. As you've made choices in how you respond to each situation, you've chosen your "avatar."* What activities and subjects did you focus on in school? Did you get a lot of education or go into the workforce young? How did you respond when someone close to you died? What did you do in the face of personal setbacks? What did you sacrifice in order to follow your values? Like my decision to stop investing in music and take a job at a tech company, these choices define a certain avatar and can have a substantial impact on who you think of yourself as.

However, each of these different avatars we could choose provides us with a common curriculum. If the synchronicities of life emerge to present us with the lessons we need, these lessons can come in any form that our life takes. There is no wrong path. Each path contains a whole life. Any avatar you choose will provide opportunities for growth and healing that are relevant to who you are specifically.

Reflecting now on my experience in business, I can see how that experience shaped who I am in ways that were necessary for who I wanted to become. I became better at collaborating with others and seeing a project through. I gained perspective on how to bring value to others. These are experiences that made me a more mature person, and have made me more successful as a musician, even though I had to leave

---

* An *avatar* is a figure used to represent a particular person in online forums or games. The user can select their characteristics, such as physical appearance, skills, and weaknesses.

it behind for a while. I suspect that, had I stayed on the music path in 2008, I would have been given different experiences that provided the same life education. In any avatar, there are ways to express the same core capacities that we were born to express. Our avatar is just the way we choose to dress ourselves up.

In the next chapter, we will explore a new way of describing time that better matches this notion of chiseling away from wholeness. Instead of seeing time moment by moment, like adding pixels to a blank screen, we are surfing a timeline from the past into the future. The holism of time emphasizes a connectedness between past, present, and future that will affect the way we think about choice.

5

TIME IS A FILMSTRIP

We think of time in very human terms. All we really experience is the present moment, so we think of time as something happening right now. We think of space the same way. It is natural to think that everything can be reduced to particles or pixels, because we can see them right here in front of our eyes. But in pattern space, we use patterns in space instead of pixels in space. Could the same be true for time as well?

A filmstrip is a pattern across time. It is a line through time that stands up together as-a-whole. You cannot reduce a film to its pieces and still have a film. We can broaden our picture of time from a moment-by-moment ticking of the clock to an adventure extended between the past and the future, from the known to the unknown.

Like chapters 2 and 3, this chapter explores some interesting and powerful ideas that can affect the way we see our entire physical world. But as before, the thread will not be lost if you skim quickly to the next chapter. There we will see how the filters in the last chapter combine with the filmstrip model of time in this chapter to help us peel away some of the filters that cloud our vision and make choices hard.

Not Moments but Timelines

What is a timeline? Think of a tennis ball flying through the air. Usually we picture its path as a bunch of snapshots that combine to describe the overall motion. Time can be broken down into pieces.

Zeno of Elea, a Greek philosopher, pointed out that because none of the snapshots can convey any clue about the direction of motion of our ball, they cannot possibly tell us whether the ball is moving. It is impossible to get "motion" out of snapshots. This is one of Zeno's famous paradoxes.[1]

By contrast, in the Holographic Paradigm we describe the tennis ball over a *span of time*. This could mean looking at the three-second span of time when it bounces across the court, or it could mean looking at the entire lifetime of the ball, from its day of manufacture to the day it gets lost under the living room couch. This moment when you are watching it fly across the court is like watching a certain frame in that larger filmstrip.

Checking Things Off a List

Story contributed by Emily Harman

I like to check things off my to-do list, and I get frustrated when some tasks linger. But I often find there is a reason. As I was developing my consulting business, my coach tasked me to create a roadmap that would resonate with my clients. I didn't have enough of a contact list to accomplish this effectively, so I was stressed about not having it complete for our next meeting. The meeting came and went without having the roadmap completed, but the following week I had a sudden 35 percent jump in membership. I now had enough data to make a sensible roadmap and felt very good about the result. The task had its own timing, separate from the part of me that just wanted it off my list.

Picturing life as a timeline or filmstrip is the most important shift in perspective required by the Holographic Paradigm. It is an interesting philosophy, but it seems a little abstract. What does it mean, from a practical perspective, to talk about the whole history of the tennis ball? This is where the virtual world of *The Matrix* becomes a handy model. If the world is composed of timelines rather than individual moments, this makes it possible for histories to be flexible. It allows for the sequence of choices that the Agents *have already made* to adjust to the free choices made by Neo and team. This is the phenomenon I call *retroactive event determination*.

To picture this better, consider the Live Photo feature on an iPhone. Live Photo takes a 1.5-second video each time you take a photo. Then, when you see the photo, you get just a hint of animation to it. It is not just a single snapshot but a moving rendition of your subject. This is also the way photographs appear in the Harry Potter movies. The photograph of Sirius Black on the front page of *The Daily Prophet* in *The Prisoner of Azkaban*, for instance, shows a 5-second loop of him screaming maniacally. The effect elicits a deeper emotion in the viewer than a still snapshot would alone.

Let's say the Live Photo feature was a bit longer, about three seconds, so it lasts about as long as it takes for the tennis ball to travel from my racket to your side of the court. Let's include in our scene not just the tennis ball but also the court, the net, and the trees hanging overhead. All these things are captured in the Live Photo.

Here a new question arises. Once we click the shutter to start the Live Photo, is the ball's path chosen in advance? This three-second window contains many distinct possibilities. The ball can hit the net, or come over to your court, or hit a twig on the ground. Within a three-second window of time, life can change dramatically. A train door may close and leave us behind, or may reopen and allow us on board. An oncoming car may correct its motion and avoid a collision with our car, or it may not. A lot can happen in three seconds! Is the story told by this three-second Live Photo already determined from the beginning?

Or is it flexible? The fate of the ball, the court, and the trees all hang in the balance.

We can get a feel for this with a story shared by a friend named Kato in Germany, who was looking for a job to bring in some much-needed income. He applied to many companies, not being very picky, and was quickly offered a job, neither his dream job nor close to home. As he was packing to leave he said, "I wish the phone would ring and someone would tell me I needn't go to Frankfurt!" Indeed, in that moment it did. Another company, more appealing than the first in terms of type of work and location, was calling to offer him a job. It turned out that the first-choice candidate had changed plans and declined the company's offer. Kato was their second-choice candidate, and the job was his if he wanted it!

In this small window of time, Kato's fortune changed dramatically. Whether the phone call came in time to reach him before he left, or whether the call got lost in voicemail, or whether the first-choice candidate remained interested in the job—all of these were possibilities that may have led to a vastly different outcome. How do we picture events like this in our lives? I often find myself thinking of the phone call he received as a lucky snapshot in time. I almost feel a sense of desperation, hoping that a moment like that will come. Much of the time it doesn't, so it feels like it is just a throw of the dice.

But if we imagine this call as part of a timeline, there comes a sense of inevitability to it. It's not that the outcome is guaranteed—we are climbing a tree and can suddenly find ourselves on a branch that doesn't include our intended result. Yet the timeline view removes the desperate uncertainty that comes when we feel like our opportunity won't arrive. We need to take action to get on a timeline, and we need to choose carefully to remain on the timeline, but we can relax a little bit within the flow of the timeline. We have built momentum to be where we are in the tree. There are important choice-points when your attention and courage are required to stay on a timeline that matches your values, but on the path itself there is some certainty to it. We are not subject to a ruthless rolling of the dice in every moment. We are on a path. Life is not an accident.

In *Living in Flow*, I presented the idea of symbolic momentum. If that model is correct, when we take actions on a path we weed out the timelines that don't match the direction we are heading. When we act, we do so with an intention or an anticipation of an experience we want to have. Gradually we find ourselves on a filmstrip that is more likely to reach that experience. Most of all, this view can increase the ease with which we make decisions and the confidence we have in our chosen direction.

To analyze more carefully how this works, let's return to the tennis court. Imagine that within this three-second window a small stick may fall off an overhead tree and onto the court, landing directly in the path of the tennis ball. If this happens, the ball's bounce is catawampus and you have no hope of getting to it. If it doesn't happen, you can hit the ball back to me and the rally stays alive. These are two distinct timelines, two possible Live Photos. Before the ball bounces, you exist in both of them. In one of those filmstrips the small piece of stick will be sitting where the ball bounces, sending it flying off course. In the other filmstrip the stick isn't there, and the ball arrives right where you expected, ripe for hitting.

Consider the branch hanging over the court. Is it predetermined whether the tree at that moment will let the stick fall? After all, there are many things that must occur leading up to the stick falling. A complex chain of chemical and biological processes must occur inside the tree branch for a stick to weaken and drop. But the whole filmstrip exists at once. If the ball bounces awry, then we also learn something about the history of the biology and chemistry in the stick. The history of the tree branch and the cellulose bonds holding it together is intimately tied with the destiny of the tennis ball.

Now there are two ways we could tell this story. Let's start with the usual way. You see the ball coming at you, and suddenly it bounces askew and you can't hit it with your racket. From this you know that it must have hit a piece of debris. If you investigated carefully, you would find that there is an overhanging branch missing a twig, and you could tell that the wound on the branch where it broke off is quite recent. You could piece together what the history *actually was*. Here the history really happened a certain way and you are trying to find out what it was.

But there's another way to tell this story that is equally consistent with the facts you know. You see the ball coming at you, and you don't know whether a twig has fallen. Therefore, you don't claim that either one has *actually happened*. A moment later, when the ball bounces askance, you finally know which history *had been* true. But the changes in the tree cells that led to the twig falling *became* true upon your retroactive inspection of the facts. In this story, up until you observed the ball's bounce, the history of the twig remained open to revision.

This second story is consistent with all the same facts, and it could provide a foundation for a world like *The Matrix*. In the first story we have an "objective, external world," while in the second story we have a personal rendering for our individual viewing. In the second story, history is a whole timeline, not a collection of separate moments. All of history—of the ball, of the player, and of the tree—is determined by its conclusion. This distinction is crucial. While a history determined by its conclusion may seem counter to our common sense, I do not know of a single experimental fact that contradicts it. This—the world we live in every day—is what we should expect to see in a world described by timelines like this.

To be clear, these timelines do not have a certain duration, say three seconds, three minutes, or three years. The timeline can include whatever duration we like, just as the system can include the ball, the twig, the net, the rackets, the tree, or whatever combination of these things we'd like. The important thing is that every timeline, filmstrip, or Live Photo has a beginning and an end that are defined *together*. Any interaction between two things marks the endpoint of one timeline and the beginning of another. Interactions, or relationships, are the real essence of reality. Interactions define our measures of time.

Which filmstrip will you experience? Until the endpoint happens, the middle point of your timeline is still up in the air. Like walking along a wire fence in an open field, everything that happens prior to the end of the timeline is undetermined. The fence posts are the known facts of your life. In between these is empty, undefined time and space stringing the facts together. What happens between the posts is flexible, existing only in your imagination, a fantasy created by your filters and colored by your emotions and impressions. As you talk to people and read the news, you gather more concrete data, and the ambiguity of these gaps turns into a fixed reality. As you make choices, you impress your own inner thoughts and feelings on the flexible time and space between the posts. Your interactions with the world—the choices you make and the relationships you form—tie you into a *particular* timeline.*

What makes the Holographic Paradigm interesting is what happens in between the fence posts. This is where the possibilities lie, the space where retroactive event determination allows your fate to hide and then suddenly emerge.

The Mechanics of Holograms

A world of timelines may be hard to imagine. Or it may be unclear why I think a world like that is so different from the way many of us picture the world. To better understand the significance of timelines, let's look at holograms a little more closely.

In college I had the opportunity to build a laser from scratch. I was enthralled with the way light can "interfere" with itself and create beautiful designs of light and dark. UC Berkeley had an air of mystique because the co-inventor of the laser, Nobel Laureate Charles Townes, was on the faculty. I could feel the history of physics as I walked through the halls. Later that semester we did another amazing experiment. We used a commercial laser (not the novice one we had built!) to create a hologram.

In the 1940s Hungarian physicist Dennis Gabor reimagined photography by using wave interference to create this new form of imaging, for which he received the 1971 Nobel Prize in physics. I was fascinated by

---

* I was given this analogy by my friend George Weissman, physicist and founding member of the Fundamental Fysiks Group, documented in David Kaiser's book *How the Hippies Saved Physics*.

the magic of it. In complete darkness, and on a table of sand completely isolated from the vibration of the floor (even the miniscule seismic rumbling of Earth would ruin the effect), we captured the light from a laser that had bounced off our target. (See Figure 5.1.) In our case the target was a simple rubber ducky, or something like that. A week later once the film was developed (it was a piece of glass with a photographic emulsion on it), I was a little surprised and disappointed to see it was just a mess of scratchy black lines. (See Figure 5.2.) Nothing special. But sure enough, when we got back into the lab in the dark and shone the laser on it, our rubber ducky seemed to be floating there in the space, near—but not exactly on—the surface of the film.*

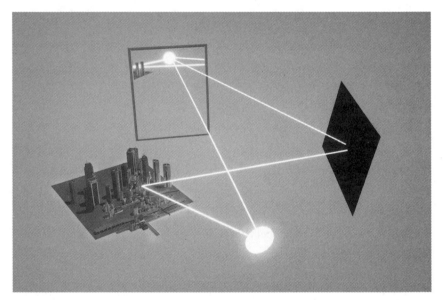

FIGURE 5.1. A setup for creating a hologram. Laser light is shone onto a model of a city, and also onto a mirror. The light reflecting off the city mixes with the light reflecting off the mirror when they meet at the black piece of film. An interference pattern is captured on the film that perfectly represents the contour of the city.

---

* We discussed the basic characteristics of a hologram in chapter 3.

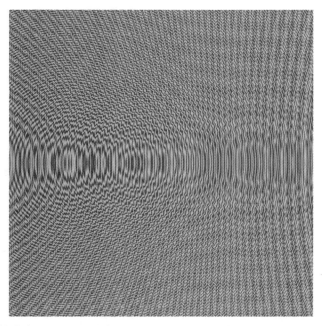

FIGURE 5.2. An example of the markings of an interference pattern on a holographic film. What you see here does not resemble the image encoded into the interference fringes, which can be revealed only under the right conditions.

I don't have that original hologram anymore—it shattered during a household move some years down the road—but in Figure 5.3 is a rendition of a hologram of a cityscape. The illusion is grand. The image does not seem to be part of the holographic film at all, but rather it appears to float in front of or behind the holographic film, like we are looking at the city through a window. When you see this effect, the science fiction of Princess Leia or Captain Picard doesn't seem like fiction anymore.

To understand a hologram, we need to understand the concept of interference. In American football, a cornerback on defense will receive a penalty of "pass interference" if they illegally touch the wide receiver on offense. The defensive player is penalized because their touch disrupts the ability of the receiver to catch the ball and is likely to change the outcome of the play.

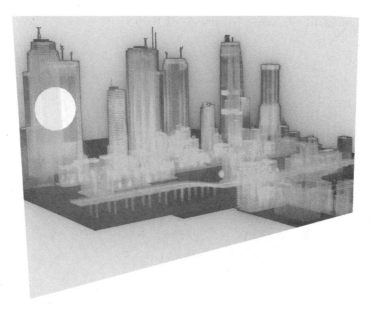

FIGURE 5.3. After we create the hologram with a laser, shining the same laser on it will make it seem like the original object is still there. A realistic three-dimensional image floats precisely where the original object was, relative to the film.

Interference of light is similar. It happens when one wave overlaps another and changes what you see. Interference applies in many other parts of nature as well. Interference patterns appear on the surface of a puddle during a rainstorm, on the surface of the ocean as the waves bounce off the shore, or even in a pair of pantyhose whose two sides rub past each other or in the mesh screen on a sliding window. Diverse and beautiful moiré patterns emerge from the bending and mixing of patterns. If you've driven by a winery or a military graveyard and watched the patterns that emerge out of the orderly rows, you will have witnessed interference. Anything that has a regular order to it will form complex patterns when it is viewed in certain ways.

So how does a hologram work? For the sake of curious readers I will go into a bit of detail.

A hologram is just a photograph that records raw interference patterns instead of washing out the patterns with a focusing lens. If we view a hologram with the same frequency of light that created it, we simply fool our eyes into thinking the city is actually there. The holographic film creates exactly the same light pattern that the actual city did. The city reappears in all its three-dimensional detail! Of course it is an illusion, but a very realistic one. You can move sideways, and your perspective on the cityscape illusion will change predictably.

Holograms illuminate the relationship between regular space and pattern space. What makes a hologram appear three-dimensional is the way in which the parts of the city at different depths slide past each other as we change perspective.

In Figure 5.4 notice the area marked by a triangle. The triangle marks the center of the piece of film itself, say at pixel (1000, 500)— that is, the 1,000th pixel from the left and the 500th from the top. Now look at the tilted film in Figure 5.5. The city clearly moves as we tilt the film in our hands.

This is a very important point. As we tilt the film through Figures 5.4, 5.5, and 5.6, the triangle is always in the middle of the film, but *it is sitting on top of a different building in each figure.** In a regular photograph the picture you see is essentially "glued" to the film, but here that isn't true because the buildings in the image move past the triangle. Some of the buildings even moved off-screen! Where did they go?! *When you look at the cityscape, you are not seeing the film itself but a pattern of light that is encoded into the film.*

Now, take another look at the background of Figures 5.4, 5.5, and 5.6. Do you notice the subtle gray ripple pattern on the film surrounding the city image in each picture? These are the patterns of light waves that were captured by the film itself. These are the interference patterns, like in Figure 5.2. Unlike the city image, they really do exist on the film, even in plain daylight. Yet they look nothing like the city that they encode. The interference pattern is "glued" to the film's surface, just like a regular photograph would be. When you tilt the hologram, the interference pattern tilts with the film because it is part of the film.

---

* The white circle is simply representing the light source, and it can be ignored.

71

FIGURE 5.4. A holographic rendition of a city looks three-dimensional. You see the city as if you are looking through a window between you and the city. The triangle marks the middle point of the film—for convenience let's say that is the 1,000th pixel from the left and the 500th pixel from the top—and contains a portion of a specific building in it.

FIGURE 5.5. Moving our vantage point to the right. The middle portion of the film, still marked by the triangle at pixel (1000, 500), now covers a different portion of the city scene. The buildings appear to have moved relative to the film!

FIGURE 5.6. Now we move our vantage point to the left. The middle portion of the film, still marked by the triangle at pixel (1000, 500), this time obscures yet another building. The key to a hologram is that the scene you see moves past each point on the film (e.g., past the triangle) as you tilt it.

The main point of this example is to illustrate the difference between our description of the *image* of the city and our description of the *patterns* on the holographic film. These are two totally different types of space! (In chapter 14 we will apply this to the whole cosmos. It will be the foundation of the holographic multiverse.) The space where the image appears—we'll call the descriptors in this space *coordinates*—is the result of an interference pattern in an underlying "holographic film" that occupies a different type of space, *pattern space*. An everyday experience, such as your commute to work, is like an image in a holographic film. But behind it are interference patterns that guide it, the pattern maps that orchestrate reality.

So far we have talked about still-image holograms. The physical shape of the cityscape was encoded into an interference pattern. In the hologram I created in college we needed to keep the object very still. But can't we, at least in principle, also encode time into a holographic form? What would a holographic movie look like? Can a single hologram represent the entire history of an object?

Holograms in Time

Our previous example illustrated what we can call *spacelessness*. Different regions of the holographic film create interference to create the illusion of a cityscape. You may know that the theory of relativity tells physicists that they should always treat time the same as space. What would it mean, then, to have a hologram *in time*? Can there be a holographic interference pattern in time as we have in space?

At first you might picture a series of regular holograms like Figures 5.4, 5.5, and 5.6 displayed in order like a movie so the city appears animated. This is like Princess Leia's avatar in the *Star Wars* scene described in chapter 3. In fact, this technology already exists. But this is not a hologram in time. A real "hologram in time" would be a single sheet of information—like the interference pattern in Figure 5.2—that contained all of history encoded within it.

This is a thought experiment. We can't do this on an actual piece of film, but the math of holographic time works just like holographic space. We are so familiar with the present moment that it is difficult to extract ourselves into a timeless perspective.

But this is just what the Live Photo description of the tennis ball flying across the court does. The ball doesn't just exist in each moment but as a whole timeline. Although you can't see the future of the ball from within the present moment, the Live Photo metaphor shows how the future is indeed part of a trajectory from the past and through the present. A hologram in time would capture this entire Live Photo animation in a single interference pattern.

If we can begin to imagine a hologram in time, and the timeline that it encodes, then we can imagine that there is not just one but many of these time holograms for each situation. *From the present moment—right now—many timelines or filmstrips are splitting off into the future like branches of a tree.* When we think about the physical world, we should think of it as *going somewhere*. Like a filmstrip, anything that happens in the moment is not just a standalone instance but is part of a *chain* of events. A timeline is like a map leading from the beginning to the end of a journey.

Let's introduce a new metaphor to make this as relatable as possible. Imagine you are riding in a self-driving car to your first day of work at a new job. You look up the directions before you leave the house and program them into the GPS. Up pops a map of your route, and as the vehicle pulls out of the driveway you kick back to enjoy your ride to work. But on the way you notice your favorite coffee shop on a side street. You attempt to turn the steering wheel but you can't! The system sticks to its preset route and won't let you change course. The autonomous vehicle is heading straight to work because the path is defined as-a-whole by the GPS from the moment you left the house.

You solve this the following morning by simply programming a new route (which includes the coffee shop) into your autonomous vehicle's GPS *before* you leave the house. Now you have a second map loaded in the system, and when it comes time to choose between them, *both maps are available to you.*

This is how a time hologram would work. Your entire journey is defined as-a-whole. You may be making a spontaneous decision to get coffee, but in doing so you are implicitly choosing a timeline. You can only get to a future that is accessible from your timeline.

Recently I attended a day-long conference in Claremont, California, where I had the chance to meet my academic crush, Mihaly Csikszentmihalyi, the author of the original book on flow.[2] Thanks to fan and friend Jennifer Redelle Carey I did indeed get to talk with Professor Csikszentmihalyi and get a photo with him face to face. I was buzzing from head to toe!

Immediately after that peak experience I used a ridesharing app on my phone to get to the airport, and I found myself riding through the familiar streets of Claremont. The streets were familiar because I had lived there twenty years earlier at a pivotal time in my life. Once I was in the car I realized I might have enough time to swing by my old address, so I mentioned it to the driver, who said hesitantly, "Umm, OK, you'll need to put the address in the mobile app." His route to the airport was already programmed into the GPS, containing the entire route from beginning to end, and he couldn't change it. In order to make a change,

I had to use the app on my phone from which I had originally ordered the ride. The information would get sent through the mobile network to a remote server and then down to his GPS device. He couldn't just turn the wheel; we had to change the route as-a-whole, in the same way that the Live Photo captured the path of the tennis ball as-a-whole.

I entered the new destination on my phone, and it updated his GPS. We were directed off the freeway, but in exiting I immediately realized there was too much traffic and I might miss my flight! Sigh. I changed my mind and went back to the app to cancel the new destination. The GPS took into consideration our current position—on the freeway exit ramp—and calculated a new, adapted route to the airport.

Each time the route was altered, the new GPS directions were consistent with the choices I had already made. When the first program led us off the freeway, the second program started with getting us back on the freeway.

The Holographic Paradigm is similar to this. When you make choices, all the possible outcomes exist as "preprogrammed possibilities" in the hologram. The present is not isolated from the future. Rather, the past, present, and future are part of timelines that exist as irreducible wholes. In the Holographic Paradigm, your present choice and the future consequences are inextricably linked together. The Holographic Paradigm is like a Choose Your Own Adventure story where you have free choice to travel through the pages of a book on one of many predesigned paths, based upon choices you make in the moment.

In the next chapter, we will further develop what happens when you start from an attitude of wholeness and use your choices to carve away and reveal a personality inside. We will explore both the inner and the outer obstacles that tend us toward smaller versions of ourselves, and the freedom that awaits us when we are able to manage our mental and emotional reactions and grow into bigger versions of ourselves.

PART 2

6

A VIRTUAL-LIKE REALITY

Having imagined life as a journey along a timeline, all of which can be encoded into a single holographic pattern, it seems natural to ask how much our world resembles a virtual reality. In the next seven chapters, we will find that it is not at all like the simplistic virtual reality we imagine in modern VR gaming, nor even the amazingly realistic one in *The Matrix*. Rather, I suspect it is a real world that is nonetheless programmed—and maybe programmable—to provide targeted opportunities to grow, heal, and adapt.

As if in an immersive training ground, from the moment of birth we are learning from our failures and our successes. We are immersed in an environment rich in thoughts and emotions, and how we navigate them affects the quality of our experiences.

But we do not usually experience the world as it really is. Our experience of life is filtered through the lens of our emotions, and we are blind to what we can't see.

My research aims to show that the cosmos is responsive. When a synchronicity occurs—someone shows up in our lives at just the right time, for instance—the vast tree of possibilities has become filtered to just one particular outcome. Our choice of action "selects" certain branches, or rather, makes them statistically more probable. Then the circumstance that occurs is likely to reflect the experiences that we anticipate. This is, at least, my speculation.

Physics then describes the world in much the same way as we would describe a virtual reality training environment. It is virtual because it is rendered on demand for you, and it is a training environment because the circumstances you experience are tailored to your choices in a way that brings you an appropriate learning experience.

I want to explore this virtual reality environment. I'll call it "virtual-reality-like," or simply "virtual-like," because I do not want to give the impression that I don't take life seriously. I think it is a mistake to say things such as, "Everything is mostly empty space, so nothing really matters anyway," or, "It's all just a game in which we create our reality," or, "The Earth will eventually get swallowed by the Sun anyway (in five billion years) so what does any of it matter?"

Reading flippant statements like these can feel invalidating to someone who is going through difficulty. Our difficulties are real. The rewards of life are real. To say it is virtual-*like* will highlight my point that there is something *not entirely real* about reality, while still honoring the validity of our experiences. We'll find that events in the world don't seem to exist independently of our witnessing of them. Yet they are as important as they can be. Subjective doesn't imply unimportant.

What are the qualities and purposes of this virtual-like world?

Do We Live in Feeling-Space?

While physicists talk of "space" as a featureless, lifeless theater, I want to emphasize a very different sort of space. I propose that we also live within a *feeling-space*.

Unlike the previous chapter, this proposal is not based on any physics that I am aware of. It's just my hunch. Take it with a grain of salt if you like.

What is this feeling-space? We are immersed in a constant barrage of feelings, both our own and those of other people. We know that we pass feelings between each other, as is obvious if you've ever tried to stay calm when your child starts getting angry at you. It takes the utmost control not to respond in anger, even if you didn't feel angry at all just a moment earlier.

Neuroscientists have identified mirror neurons as a source of this emotion-reflecting. My focus is elsewhere, on the space that is created around us by our feelings.* I argued in *Living in Flow* that the cosmos responds to what we anticipate, and this is largely a function of our feelings. So when I talk about a virtual-like world, it is by no means the dry, preprogrammed virtual reality you might immediately bring to mind. It is a testing ground for emotions. Fun, pain, excitement, embarrassment— they're all there.

Maybe the world is virtual-like so that our life can be a quest. It can be a quest for knowledge, for passion, for self-expression, for fun ... each of us has our own quest. We define for ourselves what we are learning from our lives. We all need to grow, learn, and heal. Everything we do in life is for some purpose. Even if we aimlessly procrastinate our work, our purpose is the avoidance of unpleasant feelings. We are constantly anticipating experiences we want and don't want.

Our quest changes over time. It may at one point be fully concerned with money, or love, or sex, or power, or fun, or entertainment. We may live our quest by working as much as possible, or spending as much time in nature as possible, or struggling in confrontation with others, or victimizing, or being victimized, or succeeding or failing in our finances.

Whatever the avatar we choose, the common thread is our desire to understand ourselves and heal. We may not think of it that way, but what else motivates us besides a desire to feel better? Some people are trying to feel better physically, others emotionally, mentally, or spiritually.

---

* I should note that I do not think of feelings as a physical field, and I don't expect feeling-space is measurable in the sense that regular space is measurable. It is an abstract field, like a map of the political preferences of a country or the infection rate of a disease. Emotions inside a person can quite easily be measured by neuroscientists. Feeling-space captures the concept that we can track these emotions independently from the individual people feeling them.

Lit Up by a Broken-Down Car!

Story contributed by Sky Nelson-Isaacs

I had arrived a few hours early to a training event because I was nervous about the event. Just as I arrived, my car battery died. I was worried that suddenly my spare time was to be occupied with restarting my car, and I was far from home and had to be home on time to pick up my daughter later that night. I called for road service, and the first thing they said was, "Your credit card has been declined and your membership expires today. Are you calling to renew?" I realized my luck and said yes! Then I requested a tow truck. While waiting, I decided to put away my computer and meet people. I had a couple conversations with passers-by, who both became interested in my training and decided to come. Then I had the idea to invite an acquaintance to dinner beforehand. She brought three friends who were all attending my training that evening, and we had a relaxing meal together. By the time my event started, I hadn't done any more preparation but felt very comfortable and at ease. I see lots of friendly faces in the audience, and feel loosened up from all my socializing during the day.

Life seems like a process of leveling-up (or leveling-down) our skills on a quest to grow, learn, and heal. The world is the ultimate immersive training ground, and synchronicities are the main tool for implementing the curriculum. The obstacles we face in the world are useful because they trigger obstacles within us: fear, self-doubt, self-criticism, pain, blocked grief, resentment, jealousy. To counter these obstacles, we pick up tools on the quest: self-compassion, awe, wonder, being happy with ourselves, gratitude. These are not simply ways of making life more pleasant. They are the swords and shields with which we defend ourselves against the perpetual intrusion of filters and false beliefs.

What is healing? A physical wound is an injury to living tissue that impairs the ability of the living organism to perform its necessary bodily functions. With physical wounds, we develop workarounds to cope with disability. Yet our wounds can be invisible, too. Through traumatic experiences, we can also develop workarounds—patterns of psychological, emotional, or spiritual impairment, what we call *filters*.

Living in a responsive cosmos, we have an elevated likelihood of experiencing situations that trigger doubts and fears within us. The doubts and fears are our own. The virtual-like world just presents the circumstances that trigger them.

Let's not oversimplify, however. When we experience situations outside of our control, I don't suggest we should interpret this as "on purpose." Many people experience severe trauma, which is not explained by a responsive cosmos that simply wants to teach us a lesson. Not every event is purposeful. We often cannot have clarity on the reasons behind our struggles. The purpose of living in feeling-space is to have the chance to work through these traumas and possibly heal over time.

Even as we stop being triggered by certain situations, new situations emerge. Deeper levels of hidden feelings become exposed. Or we may find ourselves overwhelmed by fears that we thought we had already dealt with, and we go back to relearn something we forgot.

Healing a layer just means removing the need for the filter so we no longer limit ourselves in the same way. Each level of growth is its own reward. We may experience joy, freedom, awe, and wonder in ways we haven't previously felt. The release of old pain is, it seems to me, the best possible experience we can have in this virtual-like world. It is a sudden leap to a better life.

Within this virtual-like framework, we live very real, substantial lives. Feeling-space makes the quest itself worthwhile.

Awe and Wonder

What is the best feeling you've ever had?

I'm not referring to your favorite physical sensations, like sexual orgasm or eating ice cream. What is the best *inner experience*? When do you feel the least uncomfortable in your skin, the least nagged by worry or self-doubt? It's a strange way to frame the question, but go with me for a second.

For me, I'm at my pinnacle when I feel awe and wonder. The experience of awe or wonder transcends any worries that might be plaguing me. When I am in awe, I am completely opened up to life. I am vulnerable but not afraid. If feelings are like colors of the rainbow, wonder is like white light. It is not a feeling itself. It is an openness to *all* feelings. When I am in that state, I really feel everything, I am not closed off. Resentment, jealousy, and embarrassment can come and go. I feel them, my eyes widen a little, and then they leave. There is no residue left. Awe and wonder are colorless.

During my first week of high school, we had a class overnight beach camping trip. I had spent middle school feeling insecure about myself and isolated from my peers, especially from girls. But on this initiation into high school, having arrived at our campsite and set up our tents, I found myself sitting in a tent chatting with a group of girls I had befriended on the hike in. I couldn't believe what was happening to me! I had built up a negative self-image from my middle school experiences, and I had no context for this experience. We became friends and gave each other massages, and they teased me about being the only boy in the tent. I felt a release of stress, the quieting of negative mental thoughts that had crowded my mind for so long.

For the rest of the day I walked around in a daze. It was not a bad, distracted sort of daze. Rather, it was like my focus had broadened. I felt spaciousness and freedom. For just a little while I was not preoccupied with where my life was going or how I would fit in. I wasn't thinking about whether I would do well in school, or how to manage difficult relationships with my parents and siblings. I was complete in that moment. Those other problems weren't resolved, but they moved right through me without resistance, and then vanished from my awareness. I was in awe. It was wonderful.

What filled me was a sense of astonished clarity. I was free of thoughts *about* myself and filled instead with thoughts *as* myself. I was caught up in a rapturous ease. Even after that moment in the tent was over, as I worked with a group to prepare dinner, I was unattached to the specifics of how we should do it. I was peacefully open-minded to different suggestions on the proper way to cook the stew or do the dishes.

I observed life gently. Rather than judging people, I gave them the benefit of the doubt. Rather than judging myself, I enjoyed being me. My thoughts were uncolored by insecurities about my appearance, my personality, or my worthiness. I saw the world directly without filters of self-evaluation, and felt welcomed by my peer group to show up just as I am. For that moment, I felt a wholeness that had been missing for the previous several years of school.

Awe and wonder are gateways to a lasting happiness. Maybe this is what we seek God for and why even a small taste of this rapture can make us Believers. When I feel awe, I feel certain of my connection to something greater than myself.

Yet when I am unhappy, it can be very hard to remember what awe and wonder feel like. When I am preoccupied with some problem in my life, it feels like the only way to feel happy is to first solve the problem. Happiness is conditional. Awe and wonder are unconditional. They allow us to feel happy even while surrounded by problems.

Rather than the feeling of relief that comes from *solving* a problem, I prefer the feeling that my problems can exist in me and there is still room to live.

When I am in awe, I feel connected to people and part of something bigger than myself. On that camping trip with my new female friends, I was welcomed into a grand experience known as friendship. I was sipping from the source of experience itself, immersed in something that people throughout history had experienced before, which was just now being presented to me.

It was on that trip that I first met Dana, who quickly became my best friend. She was the first girl I ever became close to. We could talk about anything, and often did for hours at a time. Our phone conversations were timeless and spaceless. We were unaware of how many hours it was past bedtime and oblivious to the fact that our families were just on the other side of our bedroom doors. We would dare each other to be the first one to get off the phone. As fate would have it, fourteen years later we would be married. It was an experience of grace that transcended reason.

I have felt this same awe and wonder at specific times throughout my life. My dad and I took special trips together to New York to visit his parents, and we would always drive into Manhattan to visit the American Museum of Natural History. My dad got me into geology, space, gems, and dinosaurs. I remember the vast entry way with huge, floating models of the planets in our solar system. We visited the famous Hayden Planetarium and saw a meteorite the size of a car.

I love physics because of the feeling of wonder that comes over me when I understand something for the first time. I find myself sitting with my mouth agape. I become unaware of time, unafraid of tomorrow. I am part of the beauty of the universe.

The feeling often comes from performing music as well. This is why I love improvising. The best moments happen unscripted, when I sit in with someone else on one of their songs. At a recent performance, it was when I spontaneously sat in with another performer that I felt the tingle of wonder. In that experience, I felt intimately connected to everybody in the room. We were all a part of something bigger. As I drove home in my car, I was in the same timeless state of flow that I had been in during the camping trip. My mind was peaceful.

This same experience is reported by professional teams working together to solve a problem. The awe and wonder that we feel when we work together can be the engine that propels us forward out of desperate isolation and into a world of connected problem-solving. Author, trainer, and researcher Steven Kotler studies these things in the context of flow, or optimal experience, focusing on the psychology and neuroscience behind it, and its impact on teams.

Thus, the state I am describing is nothing other than the flow state, the topic of my previous book. What makes this state possible is the removal of our filters. Awe happens when we remove a filter. We walk through the world with filters that we call "me," and when one of these is removed, we experience awe because we suddenly don't recognize ourselves. When a filter changes, the world looks very different. We lose our bearing, but it opens up new possibilities that we hadn't been aware of.

Try the following exercise if you want to see that filters lay at the foundation of the experience of flow. Imagine being at the grocery store on Friday afternoon after a long work week. You have to navigate long lines, get your groceries to the car, and then make it home through traffic. Unpleasant, right? But why? Because not only do you have to do all these time-consuming things, but they are getting in the way of what really matters to you! When I see the long lines at checkout, I imagine how badly I want to be home so that I can sit down with my family and relax.

Now imagine that you just received notification that you got a new job, a promotion, an A in a hard class, a friendly text message from your crush, or any other positive scenario that pulls you out of the moment. I bet you can now easily tolerate the inconvenient lines and frustrating traffic with a smile on your face.

These things change your experience because they address a deeper underlying problem in your life. They open you up to awe and wonder. Typically you might not even be aware of what is really stressing you out. You might have underlying angst around your crush or your job, and this angst filters out the joy from everyday situations. But when that filter is removed, even just temporarily, you are likely to endure the line with a smile on your face. You are happy to be alive.

This was my experience as a brand-new parent. Going diaper shopping the day after my daughter was born was a surreal experience. Nothing could bring me down. I wasn't *happy*, exactly, but in a sort of "dazed awe." I occupied a different world than everyone else.

Filters limit our access to joy. Then, when an event happens that removes a filter, we feel a sense of expansion. Life feels bigger than it did before.

The more we can recognize our filters, the more we can resist negativity as a habit. I am not suggesting that negativity is a problem. Habitual negativity, though, can make us miserable. When I am habitually negative, I find there is some buried feeling.

For instance, there are some musicians in my social circles who I feel are dismissive of me. It feels bad. So when they appear in my social media feed, I automatically feel negative. I may not have felt any negativity all day, and then I see their name and I feel resentful. It's a habit. It's an unhealed wound filtering my perceptions, giving me no choice but to feel bad.

We each have activities or events that allow our filters to fall away, and when this happens, we are immersed, filter-free, into the experience of being alive. When we are not in those moments, it is hard to remember that they exist. But when we are in those moments, we know firsthand what happiness and fulfillment are.

I was in the Boy Scouts in middle school, and we took a trip to a science museum called the Exploratorium. But we visited the museum at night, in order to participate in the Tactile Dome—a pitch-black, touch-only maze that you crawl through on your hands and knees. I still remember that evening, out with my friends, a long way from home, having a totally unique physical experience. It was more than entertaining. It was magnificent.

For me, awe and wonder are the opposite of dissatisfaction. They help me transcend my filters and see the world, and myself, with clear eyes. I feel stripped of my reactivity, and I am able to witness feeling-space—what it feels like to be me—firsthand.

Getting Caught Up in the Maze

The principal dilemma we are trying to gain traction on, the one that leads to feeling isolated and stuck, is the sense that no matter what we do our life won't change. We get stuck in a maze of our own creation, following the same worn tracks of emotion and thought, and we lose our ability to really communicate how we feel.

What a predicament we are in! Our filters are likely to create a version of the world that reflects what we already believe.

Consider the following two versions of a conversation:

VERSION A

Child: I'm feeling nervous about my math test.

Parent: When is it?

Child: Tomorrow.

Parent: (with edge in voice) Well, have you studied?

Child: (a little exasperated) Yes, Mom/Dad, I studied, but my teacher just gave us the review sheet today.

Parent: (firmly) OK, please go get your review sheet now and we'll go over it together.

Child: (indignant) But I am in the middle of my spelling homework. I have to do this too!

Parent: Come on, your math test is important, and you've spent the whole afternoon on this other stuff. I didn't realize you had a math test. Let's get to it!

Child: Fine! (Stomps off to get the review sheet.)

VERSION B

Child: I'm feeling nervous about my math test.

Parent: When is it?

Child: Tomorrow.

Parent: (gently) Is there something I can do to help?

Child: (calmly) No, I just need to go over it again. The teacher just gave us the review sheet today. But I have to finish my spelling first. Maybe you could help quiz me on that?

Parent: (curiously) Sure, you want me to just read you the words?

Child: (confidently) Yeah, thanks.

Consider the emotional narrative of these two versions of the story. The difference lies in the way the parent handles their emotional response.

In the first, the parent seems to be feeling anxiety around the future ("Well, have you studied?") probably because their mind is imagining the worst outcome. Within their comment is the felt expression of their fear. They are not confident that the child will figure it out, and the child feels this. Now the child's anxiety increases too. ("But I am in the middle of my spelling homework. I have to do this too!")

The parent's reaction is out of sync with the situation. Their fear, with some anger in their voice, is expressing hidden fears that the child doesn't know about. Maybe the parent is having unpleasant memories about their own math tests in school. The result is that the child is caught off guard by fear and anger that they don't understand. They interpret the words to mean that the parent doesn't have faith in them, instead of what the parent really feels—sadness over their own memories. The child feels hurt and has an emotional reaction. This is a new layer of filter.

In the second dialogue, the parent may still feel those same things but is able to control their fear. Their response empowers the child ("How can I help?") by implying that the child is capable of being in charge. By controlling their own fear, the parent is widening the scope of information they can receive. They see the fact that the child will be more likely to do well in the long run if they feel responsible and confident. Through properly managing their own filters, the parent has strengthened the confidence of their child.

This is not a prescription for handling math tests but an illustration of what happens when our feelings are out of sync with the situation. When we have hidden feelings, they can lead us into a maze of incorrect interpretations and false projections, within which we can easily get "lost." When we are unaware of our filters, we pass them on to others. In the first situation, the parent feels compelled to say, "Well, have you studied?" because their perceptions are filtered by fear. Then the kid feels criticized and forms their own filter of fear. Fear is infectious, and so is confidence!

In *Living in Flow* I emphasize that synchronicities are not positive events but neutral ones. The process of meaningful history selection means the cosmos is responsive to us, so that what we experience in life reflects our unconscious beliefs. Synchronicities can reflect our fears just as much as they can reflect our hopes. If we feel bad about ourselves, we are anticipating that we will fail, and situations become more likely that make it so.

It is important to say, and I will bring it up many times, that this doesn't mean we create our circumstances from scratch. We inherit many circumstances we cannot control, and we are not responsible for the situations thrust upon us. We will focus here on just the simplest cases: to what extent do our filters of the world spread to others and influence our own opportunities for change and growth? The influence our actions have on the opportunities or obstacles we receive is not a big one. We experience many things over which we have no influence. Yet if the effect is there at all, it may have an important effect that is worth being aware of.

This aspect of synchronicity is what makes our virtual-like world like a training ground. Meaningful history selection has a *purpose*. When we experience circumstances that reflect our beliefs and filters, it can cause us pain. But the pain helps us see the truth inside of us. When we see the truth inside of us, we become free.

Synchronicities in our lives constantly feed our weaknesses. As we wander through our inner maze, confused and misled by our feelings, we experience more and more of the frustrating or painful consequences of our flawed thinking. Yet by doing so, we can come to recognize those thoughts and change them.

When I was in middle school, I was insecure. I had warts on my fingers, I bit my nails, I had zits, and I was skinny and dorky. I believed that others couldn't really like me. My beliefs about myself were fertile ground for negative experiences to take root. I was bullied a little bit and, maybe even more painfully, I was excluded from the cool group.

When I finally made it to high school, I didn't even have to wait a week to experience an opportunity to think about myself differently. That camping trip where I met Dana dramatically changed my self-perception.

In finding our way out of the maze, we may heal parts of ourselves that became splintered off long ago.

In Hinduism this maze is called *Maya*—"illusion" in the ancient Indian language of Sanskrit. The ancients realized that the world seemed like an illusion. Things were not quite what they appeared. You are probably familiar with the experience of *thinking* you are being criticized when the person had no such intention. If you are out with a group and your friend quietly points out that you have mustard on your chin, it may feel like an embarrassing criticism, but they're doing it to help you.

This illusion dilemma probably dates as far back as human beings had any sense of self-awareness. We don't tend to interpret the world correctly. Here we are saying that this illusory veil is built not only into human psychology but into space and time itself. Things are not what they seem.

According to ancient theory, this is not a flaw but a feature of the cosmos. Swami Krishnananda writes, "Maya helps the soul to get freed from ignorance and thus be liberated from (the cycle of perpetual rebirth)."[1] Through the suffering I experience, which is influenced to some degree by my own actions, I come to see my role in things, how my choices affect each situation. I learn how to identify my hidden feelings and cease creating more trauma for myself and others.

In another writing Krishnananda says, "Maya is the cosmic aspect of the power that hides Reality's essence."[2] There is something about the hiddenness of "Reality's essence" that is crucial for helping us understand who we are.

In the Yoga Sutras of Patanjali, the filters are known as *vrittis*. As interpreted by Swami Krishnananda,

The vrittis of the mind are the powerful tendencies of the mind to move outward in the direction of objects…. (A vritti is) a modification, a shape that the mind takes in respect of a given object outside. The intention of the activity of the senses is the identification of consciousness with the object so that the consciousness may go and impinge upon the object, identify itself spatially and temporally with the object, cling to the object and imagine that its comfort, joy and delight are in the object. This is what the senses are intending to do, and they have no other activity ... These vrittis are multifarious, multifaceted, diverse, and very powerful. They are powerful because they are charged with the force of consciousness itself, the power of the mind itself ... These tendencies are so powerful that as long as they are active, there is no chance of the mind thinking in another direction.[3]

The vritti is "a shape that the mind takes," which is just how we described a filter. When my middle school filters kick in, I am likely to interpret a neutral comment or a sideways glance from another kid as critical or snide. I don't know how they really feel, but I make an assumption. The vrittis cling to what I believe, and I feel bad about myself. Maya reinforces the misperceptions and inaccuracies of thought that already existed.

Maybe the choices we make in response to our filters can be thought of in terms of the ancient Hindu theory of *karma*. Interpreted here by Paramahansa Yogananda, the Yoga Sutras say:

> One's present condition and circumstances are a composite of current free-will-initiated action and the bondage of the accumulated effects of past actions, the causes of which have often been long-since forgotten or disassociated from the results.[4]

The word "bondage" can be compared to the compulsion we feel when under the influence of our filters or vrittis. When my spouse asks if I've finished the taxes, even though our relationship may be strong, I may be compelled to interpret the comment as a threat. I say, "I've been too busy! At least I didn't spend all yesterday afternoon out with friends like you did!" which hurts her feelings. The filter I hear her through is "so powerful that ... there is no chance of the mind thinking in another direction." This is the very foundation of our personality. It's how we operate. To rip away our filters quickly like a Band-Aid would undermine our ability to function in the world.

At a recent lecture, there was a woman sitting in the center of the second row, staring up at me with arms folded on her chest and a frown on her face. I thought to myself, "This person is hating my presentation."

I start spinning out. I imagined that if she's bored, everybody must be bored. I tried to think of what to do. Be more funny? Switch to a different activity? No, I mustn't react! Just stay the course.

After the event was over, she passed me by on her way out and said, "Great lecture! And I loved reading your book." Wait, what? How did I miss that? I had been so sure that she was unhappy. I nearly lost track of my presentation in my rush to try to please her.

She didn't know how her body language was affecting me. We are constantly picking at each other's scabs without having any idea we are doing it. We perpetuate our pain by misinterpreting our experiences. We construe events as criticism rather than an opportunity to heal, and we serve as unintentional sources of pain for others through our casual comments. But if we see this as a virtual-like training ground, then these are precisely the negative messages that we need to hear in order to grow stronger. They help us discover the flaws in our thinking and liberate us from the illusion of our reactions.

We are led into the maze by the doubts we hold. There are many tools for getting out of the maze, some of which we'll discuss shortly. Seeing the world as a training ground can help us take it less personally and help us get back into flow.

7

FEELINGS BECOME
THE REALITY

Tonight I made dinner to celebrate my daughter Ellie's last day of fifth grade. I was listening to a podcast while doing the dinner preparation, so I was a bit distracted. When Dana and Ellie sat down at the table, I brought them their plates. I was worried that the food was overcooked, and I was tempted to mention it and ask their opinion.

Instead, I stayed silent. I sat down to eat with a smile on my face. Five minutes later I had heard no complaints. I looked over to Dana and asked, "Is it good?" She said, "Yes, delicious, thanks."

In choosing to remain silent instead of indirectly fish for a compliment, I shaped the feelings in that moment. If I had preemptively mentioned that the food might be overcooked, maybe Dana would say, "It's a little bit overdone, but it's fine." Then I would be disappointed because this is not the answer I wanted. Maybe Dana realizes the impact of her words, and now she feels bad for saying it.

What I say makes all the difference in the world. Maya is an illusion, but in what sense? Feeling-space is malleable. Those feelings we indulge are the ones that are felt by others and spread around us. If they weren't already true, they become true. If I approach our dinner conversation from insecurity, pretty soon everybody feels insecure. Dana is worried that she said something wrong, and Ellie is trying to help avoid an argument between us.

We evolve with our reality, and our reality evolves with us.

According to the theory of meaningful history selection I proposed in *Living in Flow*, when we anticipate having a certain experience, apples grow on those branches of the tree of possibilities. When we feel bad about something, that is the same as anticipating the outcome we feel bad about. The likelihood of experiencing a certain branch on our tree is influenced by those apples. Anticipating experiences is something we do every day, whether we know it or not. It often reflects our unconscious filters.

When we feel bad about ourselves, we identify with the version of ourselves that feels resentful, disappointed, or ashamed. When our inner critic is loud, we close our hearts in self-protection. This is anticipating the negative situation. In *The Untethered Soul*, Michael Singer writes, "You will get to a point in your growth where you understand that if you protect yourself, you will never be free."[1] You will never be free of the patterns that guide your choices.

Self-criticism can come at a tremendous cost to us. If a parent yells at us for stating our opinion, we may conclude that sharing our opinion threatens our safety, and so we stop speaking our mind around this person. We create a filter for our behavior. But when another adult who is not our parent yells at us, this filter still decides that the appropriate response is to silence ourselves. "Silencing ourselves" becomes who we are. The weather becomes the climate.

Through the mental programming of self-criticism, we gradually slice off parts of ourselves. Our wholeness becomes incomplete as parts of us no longer show themselves. *Wholeness is not so much about what is inside of us, but about what we allow to come out of us.* It's all in there somewhere, but we may avoid expressing ourselves if we feel unsafe.

Bowen Marshall discusses the personality filtering process as it is experienced in the workplace by people who identify as LGBTQI.

The meaning-making filter ... becomes a layer between the macro and micro levels moderating the extent to which macro contexts impact a person's interpretation and experiences of their micro level social identities. Specific identities become salient when a person experiences marginalization or discrimination because of the larger, social values placed on these identities ... The less developed this meaning-making filter is, the more a person allows his/her/hirs social identities to be externally defined. As the filter becomes more sophisticated, a person is better able to interpret, redefine, and ultimately self-author his/her/hirs social identities.[2]

Through self-criticism, we condition ourselves. We splinter our behavior into inauthentic reactions to the various circumstances of life, and although we have all the pieces inside, what comes out is a disjointed and incomplete representation of ourselves. How can we feel truly connected when parts of ourselves are unable to be expressed? The splintering of our wholeness leads us to feel alone.

Singer says, "Real spiritual growth happens when there is only one of you inside. There's not a part that's scared and another part that's protecting the part that's scared. All parts are unified. Because there's no part of you that you're not willing to see, the mind is no longer divided into the conscious and the unconscious."[3]

Our Insecurity Propagates Our Pain

From self-doubt come our insecurities. You might think of your insecurities as ways in which you feel hurt by the world. But insecurities can also be a way in which we hurt others. When we are insecure, we are likely to say or do hurtful things. If I were to fish for a compliment from Dana on the dinner I cooked, I might end up making her feel bad for saying the wrong thing. Although we think our "playing small" only hurts ourselves, pretending to be smaller than we are can also hurt others.

If I have a flaw, I feel insecure. But then my insecurities cause me to do things that create problems. For instance, when I go to the doctor for a hurt wrist, I might overexplain my situation, adding my own diagnosis. I may cause the doctor to become more concerned than necessary.

Before I know it, I have another appointment for follow-up labs, when all I needed was some rest and ibuprofen. Or I have to cancel a date with my wife because of a work deadline, and I feel so bad that I overexplain, emphasizing the importance of my work and thereby diminishing the importance of our date. At first she was understanding of the situation, but now I have hurt her feelings. My insecurity takes a difficult situation from manageable to crisis mode. The only flaw, it seems, is that I am worried about my flaws.

The Courage to Fail

Story contributed by Ritu Karshal

I received an opportunity to lead my first project with a small team at an instructional design firm, and because of a recent reorganization my manager was occupied elsewhere and didn't support me very much. I was nonassertive but a "good worker," and I managed the project well, for a while. But I was a people pleaser and allowed the client to gradually increase the scope of the project. Ultimately I didn't meet project deadlines even after working extremely hard and trying to do the extra work myself. I Failed with a capital F. I had been so scared of failing. That fear had often made me play small and avoid risks. Yet this terrible thing happened to me, and I found that nothing really bad happened even if I did fail. Nobody criticized me. I wasn't fired. I found that I could handle it emotionally, and I found that I had the capacity to work harder than I realized. Ultimately the confidence I gained contributed to my courage years later to move to the United States and start a new life there. It has led me to be stronger and yet more vulnerable as a writer. Working hard and failing was a gateway to who I am today: persistent, hard-working, and confident I will be OK under any circumstances.

I have seen this play out especially poorly when I am in a position of power or influence. When I am insecure, I may doubt my power or feel threatened when I am not. I am likely to exaggerate myself. I may say something disrespectful or be overly controlling in a group decision. This can undermine trust. As a member of a powerful social group—I am Caucasian and male—I can have a big positive impact by developing an understanding of these patterns and taking responsibility for my unconscious behavior, and not doing so can be very hurtful.

There is a grassroots community gathering near me that I have attended a few times. The community is somewhat like a church but focused on social justice issues. The first gathering I attended was led by a Native American speaker who talked about the importance of helping kids in school stay connected to Earth. The audience was diverse, including many youths as well as Elders, women, men, POC (people of color), and non-POC like me.

In the gathering that day, the speaker took questions. I raised my hand first. Yet as I spoke I became aware of another dynamic. I have privilege that comes from being a white man in the United States. I feel comfortable standing out and sharing my opinion because my culture supports that. I believe that my ideas are respected because that message is reinforced in the subtle wrinkles of our social fabric. I receive validation for this on a daily basis from mainstream culture, probably more so than a person of color or a woman.

But as Sky, I am an insecure person. I often doubt myself at key moments, and I often yearn for external validation to make me feel welcome or respected. As I spoke, I realized the disconnect between how I think of myself and how others see me. In asking a question, I was trying to overcome my own fear. But in their eyes, I was not an insecure little kid. I was an adult, white, straight male who appeared confident. I was suddenly aware that from the outside I may have been perceived differently than I felt on the inside, and my assertiveness may have felt domineering to others. Maybe they were right? Maybe I was domineering?

I have just as much shame and insecurity as anyone else, but my impact is enhanced because of my position in society. If I speak over someone who is systemically disadvantaged, I reinforce their pain. My insecurity has a disproportionate ability to hurt others.

I don't know for sure how to navigate these issues, but by paying attention to my insecurities I am gradually learning. For me, becoming more aware of what may be really going on in the room around me, and in other people's lives, helps me make wise choices. We coexist in a whole community of filters, a shared feeling-space. When I am sensitive to this, I find a path that allows me to be heard as I need to be, without limiting anybody else's ability to be heard. I find flow. My insecurities are valid, but they exist within a bigger context that includes many other people. I have privileges and assets that others lack, and when I am insensitive to this, my actions can be hurtful. My words are amplified by my place in the culture.

My insecurities propagate outward and my filters re-create themselves in others. They were passed to me from my parents and others who influenced me. My fear of being silenced compels me to be overly assertive and causes someone else to silence themselves. My filter becomes theirs. It is like we are inhabited by parasites that are not really a part of us. These "personality viruses" are highly contagious, getting passed from generation to generation.

Misdirected Anger Causes Wounds

Aristotle said, "Anybody can become angry—that is easy … but to be angry with … the right person, and to the right amount, and at the right time, and for the right purpose, and in the right way—this is not within everybody's power and is not easy; so that to do these things properly is rare, praiseworthy, and noble."[4]

Anger can be a very powerful tool, but *misplaced* anger can wound us and create filters that keep us from wholeness. Maybe I am angry at my spouse for not emptying the dishwasher, but my anger is covering over my vulnerability. I have a deadline for work that is making me anxious, and I feel like she is not taking me seriously. If I didn't feel scared of my work deadline, it seems doubtful that I would get angry about the dishwasher.

If I lash out at her unjustly, I'll probably cause her to say something like, "These were all your dishes in the first place, and I loaded them into the dishwasher for you!" Suddenly the tables are turned and I am on the defensive. She becomes a little bit less open to me and less likely to do me a favor in the future. My misdirected anger has created new pain for her and generated a new filter.

If instead I acknowledge that my anger is really directed toward someone at work, I build trust and connection. I could say, "When you didn't empty the dishwasher, I felt I had to do it, and I started to get really anxious because my manager expects a perfect job on this project due tomorrow, and I feel really nervous about pleasing them." My spouse will probably empathize with me and ask how she can help rather than become defensive. No filter is generated.

Or maybe I really *am* angry at my spouse, but it's for something else. Maybe she flirted with somebody at a party last week, and I am expressing my feelings through the dishwasher incident. Instead, I could say, "Can we talk about what happened at the party? I realize I am feeling angry about petty things because I'm feeling hurt." If she doesn't understand the real source of my anger, it will not make sense to her, and she'll have to pivot by creating a story around it. This is her new filter. But by identifying the real cause of my anger, she doesn't feel unjustly attacked, and my vulnerability may give her the opportunity to remove a filter instead of adding one.

When we express ourselves unjustly, we isolate ourselves. When we express ourselves fairly, we elicit empathy and strengthen connections.

Misdirected anger seems to have an intergenerational aspect as well. When we experience unjust anger from our parents, we are wounded by their anger and we defend ourselves by developing filters. We limit what we show our parents. For instance, I know that I was a sensitive kid, and when my dad got angry with me for being sensitive, I learned to hide my feelings. But I was sure that one day, when I was a parent, I would honor my kid's feelings more.

But in becoming a father myself, I have found that I am impatient when my kid is emotional. I repeat some of my dad's behaviors. When my kid expresses emotions, my sensitivity to my own dad gets triggered. I don't want him to get mad at me for being weak, so I turn around and get mad at my kid to get her to toughen up. Just as I did with my father, she learns to hide parts of herself in order to feel safe around me.

Filters due to self-criticism are passed down from generation to generation. My parents' *cause* becomes my *effect*, and in turn I become my child's *cause*. The very filters we develop to protect ourselves from those who victimize us become the lens through which we impact others.

In her book *The Politics of Trauma*, Staci Haines says, "Healing intergenerational wounds, finding new ways of addressing conflict, accountability, apologies, and amends can have a huge affect on an intimate network and family system, culture, and practices."[5]

We are all on a journey of healing. Each time we interact with people we may be participating in either the creation of a wound or the healing of one. Every choice we make is informed by our wounds and our inherited beliefs. As we'll see in the next chapter, understanding this sets us up for the potential of healing. But we are not there quite yet.

Fear Is Malleable

As I described in the last chapter, when I was in middle school I experienced bullying. It was a painful situation and some of the most unpleasant years of my life. I wasn't bullied in an extreme sense. Others have suffered far more than I did. But the experience was formative for me.

A boy named Kyle made many tough comments, backed up by the occasional throwing against a locker. I don't think he ever once hit me. But I could imagine it. Even without evidence I would back down and submit myself to him. Other times I was teased mercilessly. Kaylen found that when he called me "Weenie," I got riled up and would sometimes chase him to get him to stop. This gave him power over me, and he played that card frequently.

These were self-perpetuating cycles that had power over me because I felt powerless. If we live in a virtual-like training ground, it makes sense that its function would be to help us grow, heal, and accomplish more. To do so involves facing what we fear. It is reasonable then that the experience of synchronicity would often bring us into contact with what we fear. It is not doing so to make us feel bad about ourselves. It's doing this so we will gradually learn to be our biggest selves and fight for what we believe in, instead of becoming smaller and surrendering to fear. The training ground is helping us overcome self-doubt. These types of experiences are not failures in us but direct evidence that we are on a deeply spiritual path.

This is not to say that everything is magically good. Life is made of real problems. Identifying our fearful interpretations helps us identify where we need to grow.

Michael Singer says, "One should view their spiritual work as learning to live life without stress, problems, fear, or melodrama. There really is no reason for tension or problems … Stress only happens when you resist life's events. If you are neither pushing life away, nor pulling it toward you, then you are not creating any resistance … In this state you are just witnessing and experiencing the events of life taking place."[6] But I disagree with this, at least on its face. Tension and problems are a part of life. I don't think the path of just "going with the flow" is a spiritual way of being. If we create no resistance, we give our power away. If a person is being treated unfairly, we should fight for justice. If people are being robbed by a corrupt government, we should fight for our rights. Problems arise to show us where healing is needed, and that requires action.

So why does Singer tell us to "not resist"? He says, "What we're really resisting is the experience of the event passing through us. We don't want it affecting us inside. We know it is going to make mental and emotional impressions that will not fit with what's already in there. So we assert the force of will against the influence of the event in an attempt to stop it from passing through our hearts and minds."[7] He is saying we can fight for things, but don't fight *with ourselves*. Don't react to the inner voice that questions your worth, your skill, your qualification. Allow those inner ripples to pass over you.

Can we be free of fear? I don't think so. But we can become gradually less compelled by the experience of fear. Then we become channels for peace. We become a source through which tenderness and vulnerability enter the world. We become more effective in our work and in our relationships. I still feel fear every day, but I have experienced fewer unnecessary fearful reactions as I have healed my filters. The good opportunities that have come to me are in proportion to the amount I have let go of fear.

The movie *Night at the Museum* provides an illustration of fear that I like.* Ben Stiller plays night watchman Larry Daley, who finds that all the museum exhibits come alive after dark. Stiller quickly learns that the museum is a dangerous place. He gets bullied and threatened by the exhibits that have come to life. The stuffed lion chases him, the miniature diorama of Wild West cowboys ties him down like he was in *Gulliver's Travels*, and the monkey steals his keys and slaps him hard in the face every chance he gets. But he learns that they are simply reflecting his own state of mind. When he is afraid, they find ways to exploit his fear.

But then a villain appears, someone who knows the secret of the museum and wants to destroy all the exhibits. Daley realizes he really loves the lion, the monkey, and the miniature people. When he is thrust into the position of defending the living museum exhibits from their potential doom, his inner courage lights a spark that completely changes the dynamic. His fear goes away and he becomes fierce. The animals and display characters fall into place behind him.

Not all fear is imagined. I don't suggest that we hide our head in the sand and fix everything through positive thinking. When we are the masters of our own interpretations, we can distinguish between real threats and our own insecurities. This is the essence of the Buddhist notion of mindfulness, discerning between the qualities actually present in a situation and the qualities we contribute through our own interpretation.

---

* Spoiler alert! There is a key plot point that I am going to give away here! Skip these two paragraphs and watch the movie if you want to find out for yourself.

Fear is a bully. It beats us up, it taunts us. Franklin D. Roosevelt's famous but somewhat cryptic statement becomes clearer in this context: "The only thing we have to fear is fear itself—nameless, unreasoning, unjustified terror, which paralyzes needed efforts to convert retreat into advance."[8] When we believe messages of fear, they affect our choices and cause us to give up on ourselves. When life pushes us around, it can make us choose to play small. Yet if we don't let them undermine our confidence in ourselves, our fears may ultimately be our friends. Like Daley's living museum, this immersive training ground aims to undermine us at every step by poking at our weaknesses. If we believe the taunts, we get sucked into a cycle of experiences colored by self-doubt, where everything seems like it is out to get us.

We fill in the facts with conclusions based upon our emotions. We decide between interpretations of reality based upon gratitude, hope, and resilience on one hand and interpretations of reality based upon pessimism, frustration, and defeatism on the other hand.

Can we always choose hope over pessimism? When there are systemic injustices, I don't think so. We are not just trying to have a rosy philosophy that is out of touch with our reality. While I suggest that we can change our experience by choosing which filters we act from, I am not sure how much this can improve lives within oppressive structures or help change those structures. What can be said is that the feelings we pay attention to are the ones that dominate our lives. We do not have control over our circumstances, but understanding our filters helps us retain control over that part of ourselves we can control.

Our narrative is written first in our heart before it shows up in front of our face. Through experiences of synchronicity, those emotions influence our future, one choice at a time. Yet in my mind there is great hope. The world is virtual-like for a reason. It is programmed to help us grow, heal, and adapt so that we will do so. In the next chapter, we start the journey back in the direction of wholeness.

8

STEPS TO WHOLENESS

We have talked about the filters that hold us back from being our whole selves. Fear, self-doubt, self-criticism, and misdirected anger are expressions of our pain. But I have also framed the virtual-like world as a place of growth and healing. We have talked about what leads to wounding, and now I want to talk about the tools that bring us back to ourselves, the part of the game that can bring resolution and healing. How do we get from fear to joy? I will share my journey in discovering self-compassion and how it has proven more powerful than every one of my habitual negative emotions.

Rather than reflecting on our most desperate and challenging problems, we can think about healing the less dramatic filters that nonetheless make our lives difficult in everyday experiences. It seems unlikely that we will find a magic bullet that just solves our big problems all at once. If patterns are present in our difficulties, the best way to find them is by starting small. As we get more familiar with tools that help us choose our filters differently in small ways, we may later find relief from the bigger problems.

Leveling-Up Ourselves

I have a friend, Timothy, who faced a personal crisis during the COVID-19 quarantine. He called me one evening needing to talk because he was dealing with a difficult situation in his house. A problem had occurred with the flue in his fireplace, and smoke had come into the house. The issue was severe, and he couldn't stand to be in the house until the problem was fixed.

He was frustrated because he had felt like things were changing for him. The quarantine had provided an opportunity to practice some new ways of reacting to life. He was sick of feeling like a victim of circumstance. He was committed to taking responsibility for his attitude and feeling more in charge of his life, and he was optimistic about how he was going to be different once this experience was over.

Suddenly his fireplace malfunctioned, and he found that he couldn't sit inside his own house without feeling sick, yet he was forced to stay at home. On the phone I could feel his despair. He was disappointed in himself because he had felt like he was really changing, but suddenly the same old thoughts came rushing back in. He felt like a victim of circumstance.

Sometimes obstacles can serve as part of a healing process. Could it be that he wasn't regressing but was moving forward? Could the challenge in front of him be the next step on his path to feeling better about himself? Over the previous weeks he had learned a new way of thinking—he had accepted his quirks and limitations and felt a new sense of freedom and energy. But the learning wasn't complete. He was ready for the next level, and this required a descent into darkness.

When the fireplace broke, it led him straight into a familiar set of feelings of despair. He could now see so clearly where his enemy lay. Lurking there in his mind and body was a message just waiting for the right circumstances to emerge. "You're not physically resilient enough to live the life you want to lead," he told himself. I knew that Timothy pushed himself hard as an outdoorsman, and this behavior had gotten him into trouble before. One Halloween night a few years earlier he had taken a kayak out on the water near his house in the middle of the night. He was alone, and he didn't discover his kayak had a leak until he was far from shore. He was sinking, and on that foggy night, he was lucky that a resident in a nearby house on the water heard his cries for help.

Did this new experience with the house have anything to do with the kayak incident years ago? In one story, he pushed his physical limits to the danger point. In the other story, the feeling of unwellness caused by the odor made him feel fragile. He felt that his lack of health meant failure. Proving his vitality helped him feel competent and avoid feeling weak.

Yet the past few weeks had been about accepting himself as he was. He had made some real changes in his self-talk, accepting himself as he was. And suddenly here was a big test. The smoke in his house made him feel fragile and weak, but he didn't have to respond to that in the same way anymore.

Rather than an indication of failure, could this be a gateway to a new level of learning? Could this be the opportunity to finally shed the old habits he'd been uncovering? When we see our difficulties as part of a *quest*, we can shift the way we relate to adversity.

It's not just about a shift in perspective. It's more than trying to see his house problem as "a learning experience." I'm suggesting that this specific challenge is a synchronicity tailored to him. It is part of a curriculum that is useful to him to move to a higher level of functioning. While it was easy to see this setback as just another failure like all the rest, it was not the same. This hurdle marked the successful completion of a process. It was an indication not of what he had been doing wrong but of what he had been doing right.

As it turns out, it's funny that he reached out to me because, unbeknownst to him, I dealt with this same issue a few years ago—twice! His situation was a synchronicity for me, too. I was able to be a better friend to him and heal my own experience by empathizing with him.

Healing his negative self-talk is a way for Timothy to get to a new level on his quest. We shouldn't think of the next level as a *higher* level or *better* level. The curriculum doesn't go from high to low. It is about peeling back. We are immersed in an ocean of emotion, and each layer of reactivity that we peel back exposes another level of feeling. Some of us are dealing with obvious trauma, others dealing with less obvious sources of pain.

A Twisting Creative Path

Story contributed by Gregory Berg

I spent a month traveling in France, prompted by a recent breakup. I was traveling alone, so I looked online for groups of English-speaking travelers. On a whim, I focused my search narrowly on mindfulness and consciousness. I soon came across a group that seemed interesting, led by an American woman who had roots in my own hometown in the US! Connecting with her led to finding a cheap apartment within view of the Eiffel Tower. Our friendship strengthened upon returning to the US, where she now resides just a few miles away. Through her intuitive guidance, I came to a new understanding of my own work and created an overarching online platform that I'd unsuccessfully tried to build for years.

Peeling back layers that keep us from wholeness doesn't have a certain order. We might start by feeling anger at everyone, then come to find that underneath that is frustration with a specific area of life, then later come to see that we are scared of failing and being abandoned by our family and friends. Under that we may finally discover sadness at feeling alone, and under that may be awe and gratitude for the love we have right now.

We take these layers of experience one at a time, and each of them feels very different. Our way of experiencing the world is different at each level. Each level of experience is more raw and vulnerable than the previous, but they are not better or higher than each other. The closer they get to the center of who we are, the less limited we feel.

The point of the virtual-like world is to level-up the quality of our experience, which has a lot to do with how we feel but is not limited to that. Real change can happen in our lives as we level-up our experience. We may more easily obtain the raise we have been seeking for so long, because we stop undermining ourselves at the last minute. Or we may get the more satisfying home life we've been wanting, because we face our fear of asking our boss for time off.

When we level-up ourselves, we become more wholly ourselves. There is no highest level to get to on this journey through our virtual-like reality. It is a game we can play but can never win. Rather, we learn how to watch for wake-up clues and find a deeper sense of awe and wonder in the everyday experiences.

Synchronicity as a Process of Healing

We can use experiences of synchronicity as a means of discovering the next lesson we need to learn from life. Events that feel like synchronicities shouldn't be thought of as positive or negative but rather seen as tending toward growth. What makes it a synchronicity rather than just a coincidence is how tightly it connects to a future experience you are seeking to have.

Timothy's experience getting blindsided by the smoke in his house can be seen as part of a cycle of synchronicity that was helpful for healing. This cycle is illustrated in Figure 8.1.

The cycle assumes we have a hidden feeling, some way in which we are unconsciously wounded. First an experience happens that compels us to react—the parent's fear about their child's math test in the last chapter. We have a reaction even before we are aware of the underlying feelings, and our reaction intensifies the dynamic. The parent was not consciously aware that they were afraid for the child's future, but their comments made the child defensive. These patterns of hidden feelings may be what Hindu philosophy calls "samskaras."

Because of our enhanced discomfort, though, we feel a deep yearning to resolve the pain. More than succeeding at math, the parent wants the child to gain confidence and to feel they can trust the parent. This activates something inside of us, a desire to be healed. We think to ourselves, "I just don't know how to manage this situation. I'll do whatever it takes." As a result, we anticipate the experience of feeling better.

FIGURE 8.1. Healing is a process of bringing our hidden feelings to light. We naturally want to heal, so we naturally anticipate the experience of being healed. The impact of meaningful history selection is to increase the likelihood of experiencing circumstances that trigger our wounds and give us the opportunity to choose differently than in the past.

This is the *anticipated qualitative experience* from meaningful history selection that influences the likelihood of a synchronicity. It now become more probable that a synchronicity will occur, which gives us another shot at growth.

The synchronicity may not be a positive thing that we recognize as helpful. We find ourselves nagging the child to clean their room, and another argument happens. We should be on the lookout for this. Our hidden feeling has been triggered again—fear that the child will never learn to clean up for themselves—and we find ourselves in a similar experience to before. The details are different, but the feeling is the same.

Now we have the chance to choose differently. If we pay attention to our experiences in the cycle, we will notice the feelings and thoughts we have along the way. We may be able to analyze some of them objectively to see what's true and what's not. We may catch a glimpse of memories from our past and start to see a pattern unfolding. (Later in this chapter we'll discuss the ARGH! process, which helps figure out what pattern is buried there.) Once the pattern becomes conscious, we can now choose a different action. If we break the cycle of action-reaction, we jump to a new level. This is healing.

We can call these synchronicities, but be aware that they are not necessarily easy experiences. Are feelings of shame healed by remaining insulated from others? Are feelings of financial insecurity healed by being handed a load of cash? Are feelings of not fitting in healed by suddenly becoming famous? I don't think so.

The synchronicities we should expect are situations that lead us to reexperience some aspect of a difficult situation. As long as we keep responding to a situation with the old filter, we will experience it in the same way but more and more intensely through each recurring circumstance. As we grow to understand the pattern, one day the same circumstance will happen and we will respond differently. In that moment we have reprogrammed ourselves a little, and our trauma response lessens.

At this point, we no longer subconsciously anticipate situations that will help us heal, because we have *already* healed. Pema Chodron says,

> Nothing ever goes away until it has taught us what we need to know ... It just keeps returning with new names, forms, and manifestations until we learn whatever it has to teach us about where we are separating ourselves from reality, how we are pulling back instead of opening up, closing down instead of allowing ourselves to experience fully whatever we encounter, without hesitating or retreating into ourselves.[1]

Healing is about changing our response to a situation, guided by a new insight. Once we have been through a situation enough times, we can uncover the hidden feelings compelling us to sabotage ourselves.

Then it seems quite natural to step outside of the cycle, which ceases perpetuating itself.

One evening I was sitting around a campfire with a group of people who invited me to play some music on guitar. I know a lot of songs, and I've written plenty of my own as well, yet in that moment I couldn't come up with something to play. It's my *job* to play music for people, yet I turned down the offer.

Why? It's not that I couldn't think of anything to play, it's that I couldn't think of something *good* to play. I wanted to play, but every song I thought of made me feel ashamed.

Shortly afterward, I was driving my daughter to school and I put on a CD of Huey Lewis and the News. My daughter asked if we could change the music, and the same feeling of shame flooded through me. A memory from childhood resurfaced.

It was a cold morning in ninth grade as I was riding to school with my two older stepbrothers. We were pulling out of the driveway and I asked my brother to put on Huey Lewis and the News. I listened to Huey Lewis all the time with my dad. However, my stepbrothers had a different dad and didn't share a fondness for the music. They said something disdainful, which made me ashamed about my taste in music to this day. I can imagine the exact surroundings of that moment of filter imprint. I was sitting in the back seat, seeing our front gate pass us as we pulled away from the house. I can smell the leather seats of my brother's Volkswagen Rabbit. I still feel a twinge of shame when listening to Huey Lewis, like it's a guilty pleasure.

This came back when my daughter wanted to change the music in the car. It also made it impossible to decide which song to play at the campfire. I was ashamed about the music I liked. Staci Haines describes how our bodies hold on to memories until we are ready to heal them.

> In this somatic opening, as tissues change, so do belief systems; as emotions get to be felt, more ease and emotional range appear and fear decreases. As the soma both relaxes and enlivens, a range of emotions often appears, particularly those we were unable to have or integrate.

This may be sadness and grief, anger, fear, and/or happiness and satis-faction ... through the somatic opening processes, history that has been stored in the tissues can surface.[2]

I didn't want to feel like this anymore. Some months later I found myself sitting around another campfire and was asked to play a song. I thought of a group sing-along I had written, but immediately I felt that wave of shame. It was a powerful and vulnerable song for me, about loving the ocean. What if they were bored? What if they laughed? What if they didn't get it? But my desire to be healed had arisen, and I saw my chance. I had new perspective. I wasn't thirteen anymore. These people weren't my brothers teasing me. This was a supportive community of thoughtful and caring adults.

I stopped holding myself back and sang the song. It brought joy to me and to the group. We felt connected. A layer of shame had been removed. This single choice healed the pattern just as powerfully as the original wound had caused it.

It may not always be clear how an event is helping us heal. Some-times when synchronicity shows up to help us heal, it acts as a great disrupter. While we may not be successful in the ways that we want, the recurrence of difficulty in our lives may be a sign that we are progressing on a path of healing and growth. If we see this as a learning cycle, each time it happens we gain more clarity about our role in it, and we are in a better position than before to make a different choice.

A New Toolbox

Last year I "broke up" with my best friend. My wife had introduced me to Clarence over ten years ago, and we had quickly discovered we had a lot in common. We loved to talk about esoteric ideas and ruminate on the philosophy of life. We had big visions of ourselves, even if they were often wishful thinking, and we enjoyed our fantasies together.

At that time I was becoming more aware of my own painful patterns. I talked a big game, but when it came to implementation and really showing up when it mattered, I would often struggle. I have been a big-picture thinker my whole life, but insecurity had held me back from experiencing the apples I wanted. I desperately wanted to change.

Clarence and I shared these traits. Our conversations followed well-worn tracks. He wasn't changing, and I didn't feel I could change when I was with him. I could no longer enjoy our fantasies of the future. They made me feel nauseous. I wanted something real, and Clarence still reveled in the fantasy.

I felt increasingly dissatisfied by our interactions and scared that my own life would never move past this adolescent phase. One day I told him this and cut off communication. Our friendship was over.

My friendship with Clarence had shown a light on my own filters, helping me see myself more clearly. The experience of losing our friendship was also a mirror for me. When I felt afraid, I isolated myself. There was a missing depth in our connection. It is difficult for me to feel how much I need my relationships. I protect myself from criticism or hurt by distancing myself. I left a friendship in order to stop running away from intimacy. That irony is not lost on me!

Once again Pema Chodron writes,

> Those events and people in our lives who trigger our unresolved issues could be regarded as good news. We don't have to go hunting for anything. We don't need to try to create situations in which we reach our limit. They occur all by themselves, with clockwork regularity.[3]

I was desperate to be different, and I found myself in the thick of my patterns. I couldn't yet see who I wanted to be, but I was clear who I didn't want to be. I knew that there must be something going on inside me, underneath the visible layers, something that made this discomfort repeatedly erupt in my life. What would it look like to be different?

Soon after, I was putting the final touches on a research paper before a deadline. Suddenly I saw the project with fresh eyes. I realized that my research had gone in the wrong direction. I could make some worthwhile changes, but I needed more time. It was simply too late.

I felt despair. I was mad at myself for not leaving time for this. If only I had worked harder! I felt I had let myself and the client down, and I wanted to give up.

In that moment something flashed across my computer screen. This was August 2018, and the news was full of stories about the devastating fires in the Amazon rainforest. These were the most destructive fires the Amazon had ever seen, in my mind clearly indicative of the new era of climate change.

Suddenly I gained a wider perspective on my problems. My own mistakes felt irrelevant. The project needed to get done. My research paper had nothing to do with climate change, but my showing up fully to the challenge has everything to do with climate change. I needed to get myself out of the way and complete my project the best I could. Then I needed to move on to the next thing. I needed to let go of the inner criticism and keep doing good things in the world.

The best thing I can do to heal the wounds of climate change is to heal the wounds that stop me from contributing my best.

The devastation in the Amazon expanded my perspective. It didn't change the details of my problem, it simply helped me stop *worrying* about the problem. I stopped thinking about whether it was good enough and got back to work.

I still felt a weight of despair that made me want to crumple like wet spaghetti, but I took one step after another to keep on the path. I cooked breakfast for my family, and I still felt despair. I helped my daughter get ready for school, and the despair was there, but not as loud. By the time I sat down in my quiet office and made an outline of the steps needed to complete the project, the critic had gone silent. I was back at my baseline, ready to do the work, deadline or not.

I still felt hopeless. I was not faking optimism. The Amazon was dying, and I didn't see how I was going to complete the paper. But I had felt a shift. I had recovered my motivation to *do something*, and I was committed to being of service. I was *free*.

I had been lost in a maze of despair. My self-critic had been so loud that I couldn't hear anything else, stuck in a tunnel where all I could see was one path forward. This is what was going on underneath the surface layers, causing me to dump Clarence and nearly give up on the research paper. I had been unkind to myself and feeling miserable about everything.

But here was a different way to think about my problems. Rather than pointing my finger and looking for fault, I could be nicer to myself. If I was more compassionate with myself, I could back out of the tunnel and see the other paths available to me. Self-compassion was the tool I was looking for.

Being gentle with ourselves neutralizes the inner voice of criticism that sabotages us. Starting from kindness, I was able to think about how to respond to the situation effectively. This removed the pressure to get it just right. I wrote a brief email to the client acknowledging the delay and letting them know to expect the report early the next week. Then, with renewed vigor and confidence, I dove back into the project. I lost sleep and drank more coffee than usual, but by the middle of the following week I had transformed the paper. Little if any damage was done to the client's trust in me, and my only regret was the time wasted in self-criticism.

Being compassionate with ourselves doesn't mean being complacent. Life requires action of us. Both spiritual progress and material progress involve action. *By responding to our challenges with compassion for ourselves, we have more compassion for others too, and we become more powerful forces in the world.*

That night I had a dream. I was living in a house from my childhood, and I walked into the backyard to discover a door to a hidden basement. With curiosity I entered the basement and found a few racks of metal shelves, as if in a hardware store. On every shelf were toolboxes filled with shiny, stainless steel tools. There were wrenches, pliers, screwdrivers, and hammers. I had found buckets and buckets of new tools in my own basement!

These tools exist in me, in my own subconscious. Wrenches of tenderness, screwdrivers of self-compassion, pliers of patience, and hammers of persistence help me notice my assumptions, stories, and beliefs. These tools help me remain grounded in "not-knowing."

Even hopelessness helped me. Hopelessness opened my eyes. Hopelessness helped me feel nonattached to the outcome. It led to commitment. From a place of hopelessness, I was not in it for myself anymore. I was naked and just wanted to know how I could help make things better.

Nonattachment, or *vairagya* in the Sanskrit language, is an important concept in Hindu philosophy.[4] It comes when we detach ourselves from expectations and just do things to be helpful. There is no room left for thoughts about how we could have done it better.

In that space, we can find our core power. Everything we do is impactful to others. All hands and all voices are needed, imperfect though they may be. It is an act of selfishness to give up.

Whatever it is I can contribute, surely that is better than moping about my personal failures. Self-compassion lit a fire under me that extracted me from the maze of self-defeating thoughts and feelings.

The Unreasonable Effectiveness of Self-Compassion

Self-compassion is a highly developed concept that has been studied extensively for thousands of years, most notably in Buddhism, and can help us step out of that cycle of negative interpretation. Chodron says,

> Really communicating to the heart and being there for someone else … means accepting every aspect of ourselves, even the parts we don't like. To do this requires openness … not fixating or holding on to anything. Only in an open space where we're not all caught up in our own version of reality can we see and hear and feel who others really are, which allows us to be with them and communicate with them properly.[5]

Chodron is speaking about filters. When we don't like the feelings generated by our filters, we are tempted to push them away. But she is emphasizing the importance of remaining open when negative feelings arise. She uses the words "accepting every aspect of ourselves" because in many cases our reason for resisting our feelings is self-judgment. When we can accept the real version of who we are without judgment, including our fear, insecurity, and self-pity, then we can show up authentically and "communicate properly."

Effective communication relies on self-compassion. When we have judgments of ourselves, we might find it hard to acknowledge our flaws, and this limits what we are willing to communicate. I couldn't continue conversations with Clarence because they brought me too close to my own pain.

We color everything with our inner experience. In striving to address the *content* of the problem—increase the bank account or find the desired relationship—we are compelled to make choices from feelings of inadequacy, insecurity, resentment, or shame. These represent the *context* within which life is happening. In trying to fix the thing we are ashamed of instead of inquiring why we feel ashamed, we may do or say things we later regret.

When we feel self-compassion, we stop trying to protect ourselves. Chodron explains:

> A further sign of health is that we don't become undone by fear and trembling, but we take it as a message that it's time to stop struggling and look directly at what's threatening us. Things like disappointment and anxiety are messengers telling us that we're about to go into unknown territory.[6]

Self-compassion focuses on the filters themselves, not the content they apply to. It addresses the finger that is pointed, not what it points to. When the content seems to threaten our security—our marriage or our job seems to be at risk—it can be hard to resist obsessing about it. Self-compassion eases that worry. It provides space around us to breathe without the immediate threat of catastrophe. It is a first line of defense that helps us say, "You're OK, you're alright. Let's take a deep breath and find out what the real problem is here."

Self-compassion is a practice. Each event has the potential to lead us back to self-criticism, which may cause us to be angry, hurtful, or cold with people who love us. Self-compassion is the birthplace of compassion for others. The practice of self-compassion involves being vigilant with the layers of self-criticism that keep trying to take up residence within us; we steadily soothe ourselves just as a parent would soothe a child. We can keep coming back to self-compassion. It can counterbalance any of our critic's antics because it doesn't care about being perfect or assigning blame.

Eugene Wigner is one of my favorite physicists. He wrote a paper titled "The Unreasonable Effectiveness of Mathematics in the Natural Sciences," reminding us that it is easy to take for granted the fact that mathematics is so amazingly effective at describing the world. The better we understand the math describing an atom, for instance, the more accurately we can predict its behavior. Why is that so? "The miracle of the appropriateness of the language of mathematics to the formulation of the laws of physics is a wonderful gift which we neither understand nor deserve."[7]

Similarly I am amazed by the "unreasonable effectiveness" of self-compassion. It is the most effective strategy I have found for overcoming inner problems, because the problems can only implant in the places where we feel critical of ourselves. Of course, our problems don't disappear just because we are kind to ourselves, but we are more capable of working through them when we don't let perfectionism or blame get in our way.

This is a gift we "neither understand nor deserve." We don't have to earn self-compassion. It is not a reward for good behavior. Even with our many faults, if we are compassionate with ourselves, our hard edges will gradually become softer and our problems will ease. This is the essence of grace. We don't have to ask for permission to be compassionate with ourselves. We don't have to have a certain GPA or salary. We can choose to be compassionate with ourselves anytime. This doesn't free us from consequences—rather, it helps us fully account for our mistakes. Self-compassion brings humility and makes reconciliation possible.

Self-compassion is a *practice*. Even the act of not being compassionate gives our critic ammunition against us. "You should have been more compassionate with yourself! You'll never get it right!" How do we escape this perpetual loop of self-critique?

Self-compassion is a leap.

Self-compassion is a clean break from the cycle of feeling bad about ourselves. We don't have to wait until we have enough evidence that we are a good person, worthy of love. We will never find enough evidence for that if our filters are constantly coloring our perceptions. We don't need to understand or feel that we deserve the grace of self-forgiveness and self-appreciation. We just take it because it is there to be taken. Chodron discusses this:

> Rather than indulge or reject our experience, we can somehow let the energy of the emotion, the quality of what we're feeling, pierce us to the heart. When we reach our limit … a hardness in us will dissolve. We will be softened by the sheer force of whatever energy arises. When it's not solidified in one direction or another, that very energy pierces us to the heart, and it opens us. This is the discovery of egolessness. It's when all our usual schemes fall apart.[8]

When we soften our reactions, we give ourselves time and space to reflect. This is a leap from *content* to *context*. If we feel discomfort, *why* do we feel discomfort? What does our worry or defensiveness point to? Do we feel ashamed, inadequate, or angry? If we can catch a glimpse of the filter, such as, "I will never figure this out!" we can assure ourselves that we are OK even if we are not perfect.

The practice of self-compassion lessens the handholds that criticism can grab onto. Instead of trying to *solve* our problems, we can ask ourselves: Why we are critical of ourselves? Why are we unhappy with our own performance? What were our expectations of ourselves?

Practicing self-compassion doesn't make us money and pay our rent. It instead removes the compulsive behaviors that stand in the way of taking effective action. We are more powerful when we make mistakes and are compassionate with ourselves than when we try to be perfect and are hard on ourselves. In trying to be perfect we get frustrated and make only choppy progress, but in being compassionate we enter flow and find that people and things show up in our path to help us, even in our mistakes.

Self-compassion lubricates the gears and allows something greater than our own willpower to invest itself in us. When we feel self-compassion, we are brought back to awe and wonder.

> How we stay in the middle between indulging and repressing is by acknowledging whatever arises without judgment, letting the thoughts simply dissolve, and then going back to the openness of this very moment … Out of nowhere, we stop struggling and relax. We stop talking to ourselves and come back to the freshness of the present moment.[9]

Chodron is talking about the *leap*. We simply go back to the openness without responding to the judgment. "Out of nowhere, we … relax." Then synchronicity can show up. It doesn't judge us and say, "You are too flawed for help." It looks at our mistakes and sends a hand to help us correct our course.

In mastering flow, we make errors constantly. Through a profound sense of compassion, we manage feelings of failure and yet continue moving forward. To stay in flow requires constant self-forgiveness. The quicker we can recognize our filters and accept our flaws as they emerge, the quicker we can see a way out of our current mess. We can grieve our disappointments immediately, openly, without misplacing our grief or anger.

The maze inside our mind has many obstacles—walls of preconception, pits of judgment, gates of self-pity, and locked doors of unexpressed grief. You can find the tools of tenderness, patience, perseverance, and self-compassion in the basement underneath the maze. These tools can help you escape from negative self-talk if you get trapped by it.

Stop searching for a way out. Just sit down in the dark where you find yourself right now. Pull out a pencil, and begin writing a new story.

The ARGH! Process

None of the stories shared so far contains an experience of total enlightenment. They reflect a cyclical process of growth and change. The filters we carry inside ourselves change with each experience. In each cycle a little change occurs, like the erosion that happens on cliffs during each high-tide cycle. I may yearn to stop making mistakes, but in nature change happens through mistakes. Mistakes are necessary. I want to change all at once, but nature evolves gently and gradually. Children grow up through steady tenderness and nurturing, not by trying to leap-frog past each lesson.

When our flow is disrupted, we want to yell "ARGH!" We get frustrated with interruptions and obstacles. That simple expletive can remind us of a way through. The ARGH! process is a way to apply self-compassion to problem-solving. We can use it whenever we feel like yelling the word itself. The process is:

Accept the situation

Recognize the pattern

Growth valued over accomplishment

Heal the pattern

Be on alert for the moments when it first dawns on you that you have a problem. This could be when the problem happens, or it could be later when you come to find out about the problem. It is the moment of shock when you say, "ARGH!" to yourself.

Recall the story of the report I didn't complete on time. When I realized I would miss my deadline, being compassionate with myself meant *A*ccepting the situation—accepting that I felt like a failure. Maybe our own mistakes in judgment have led us into trouble. Maybe a client will be angry and won't renew our contract. We start by accepting these possibilities without pushing them away.

Once we have accepted the situation, then we can *R*ecognize the pattern—I had let myself get mired in unimportant details until it was too late to see the big picture. This was not the first time this had happened to me. Focusing on the pattern is a way of playing the long game rather than going for the quick win. If we haven't yet accepted our feelings of disappointment, it may be too painful to see the pattern. If we are living in fear that our business will collapse, it can be hard to acknowledge our own role in the problem. Compassion helps. Having accepted the situation, we have forgiven ourselves and others for the part each has played in the problem. Then we can more readily recognize the pattern and learn from it.

Once we have seen the pattern, we can focus on *Growth* over accomplishment. Becoming a better researcher, writer, and professional is more impactful in the long run than completing one report on time. If we cling to the accomplishment, then we might bang our head against a wall. We might write a defensive email, creating bigger problems than the ones we face today. When disruptive events occur, switching from seeking success to focusing on growth can help us adapt and thrive in the long term.

Finally, we can *Heal* the pattern. This is where we make a different choice and step out of the cycle, the point at which we remove a filter from our perception. In the case of my report, I could more clearly see the pattern of imagining the criticism that I would receive and then giving up on myself. By facing that imagined future and choosing to respond to it with a different action, I removed some of the power it had over me. In response to a familiar set of feelings—despair, frustration, hopelessness—we forge a new path through the woods.

Having a little more insight into my pattern, I recognized how I often give up because I think I have failed. Since that experience I have had other opportunities to collapse when I feel discouraged, but having found a new path through the woods, I don't get lost in the discouragement. By focusing on context over content, I can more easily find a persevering attitude when I feel despair.

Accepting a bad situation can be hard. Yet acceptance is powerful. It leads us out of inaction. Climate change is a wonderful teacher of acceptance. It accepts no quick solutions, so we are forced to confront our grief. We have no choice but to play the long game, which means valuing growth over accomplishment, relationship over progress, and feeling over action. But these are the same choices that will heal our relationships with our parents, with our colleagues, and with ourselves. Climate change provides a backdrop against which we can each individually grow and heal. I so dearly hope that we can use it in this way.

If we understand the pattern of misperceptions and misaligned choices undermining us, and we make growing out of the pattern a higher priority than solving the problem, we are on a trajectory to heal the pattern. When the situation shows up again, we will be ready to approach it differently.

The ARGH! process—Accept the feeling, Recognize the pattern, Growth valued over accomplishment, Heal the pattern—can help you choose your emotions and actions in the moments when your flow is disrupted and you are thrown off balance. In the next chapter we will see how wholeness, compassion, and a growth mindset can help us make more conscious choices toward healing our patterns.

9

HEALING AS A CHOICE

The Constant Struggle to Avoid Vulnerability

Although I have suggested that being compassionate with ourselves is a leap we can make at any time, we should not expect to change quickly. It's popular today to think we can easily embrace positive feelings and release negative feelings, but this has not been my experience. When I pretend I am OK, my frustration finds some other way to be expressed. Although it may work for others, simply repeating the mantra "I am grateful" has rarely made me *feel* more grateful. Rather, I have been through an almost surgical process of teasing apart the thoughts and emotions that cause my negative feelings. In order to change, I have needed to understand the inner source of my problems.

We get to happiness by opening to a deep well of vulnerability within us. What is vulnerability? Researcher Brené Brown says, "The definition of vulnerability is uncertainty, risk, and emotional exposure. But vulnerability is not weakness; it's our most accurate measure of courage."[1] In my view, vulnerability is powerful because it reflects the truth about our circumstance. The reality of life, if we can see past our filters, is that everything is uncertain. No matter how firm our stance is, life has ways of undermining us, whether through illness or natural disasters, through unexpected people showing up in our lives, or the loss of someone or something we care deeply about. We can adapt better to life, and therefore have more influence over the direction it takes, if we understand and honor our vulnerability. We don't have to say anything vulnerable or outwardly express vulnerability, but if we feel our vulnerability, it will show in our actions. I don't think there can be authentic spiritual experience without profound vulnerability. We can think of spirituality as nothing more than the continual practice of relating to vulnerability. Negative emotions are our natural response to feeling vulnerable, so if we try to always be positive we run the risk of disconnecting from the truth of who we are. Authentic spirituality must therefore embrace all aspects of ourselves, both the positive experiences and the negative ones. Authentic spirituality is about what it is really like to be us.

Finding a Place to Settle Down

Story contributed by Ian Griffin

I'd just returned to England after three years of study and travel in Portland, Oregon. I had no idea where I'd live, for both the town where I'd been raised and the college town I'd recently left were unappealing. I hitchhiked down to Cornwall to visit friends, and on the way back north someone suggested I visit Bristol. I followed the whim and visited. In the US I had made two friends, Faith and Martha, at the shoe store where I'd worked. After dropping my bags at the hotel in Bristol, I'd walked no more than five minutes down the street when I saw Faith walking toward me. I had no idea she was in England, let alone Bristol. It turns out she had friends there and was visiting on vacation. Our nice experience meeting in Bristol inspired me to stay there, where I settled in for a satisfying and productive couple of years.

Brown says, "The fear of failing, making mistakes, not meeting people's expectations, being criticized keeps us outside of the arena where healthy competition and striving unfolds."[2] Vulnerability allows our economic activity and our professional relationships to be informed with a sense of spirituality. It is not a touchy-feely or dogmatic sort of spirituality, but an in-the-trenches effort to improve our relationships to work, to each other, and to ourselves. Making change in the world—dealing with climate change, healing wounds of social or racial disparity, reducing unnecessary violence, improving quality of life for all—comes from each of us making wise decisions, especially in the context of our organizations. This becomes more possible if we can be vulnerable.

If we are available to them, lessons in vulnerability are available to us at any time. I have been studying spiritual teachings and had spiritual practices in my life for a long time, yet my wife will laugh and roll her eyes when I claim to be "spiritual." She sees truly—I am constantly wrestling with new layers of self-deception, new ways in which I am trying to convince myself of my worth and avoid feeling vulnerable.

We talked about the illusion of Maya in the last chapter. I'm reminded of the book series *Spy School* by Stuart Gibbs, which Ellie and I listened to in the car. Maya is like the enemy spy organization, Spyder, incessantly devious and always a few steps ahead of everyone else. In one particular plot twist (spoiler alert!) the protagonist, Ben Ripley, goes undercover with the enemy, only to find in the end that the enemy knew all along he was a double agent. They used his math genius to calculate the trajectories for a missile attack. He thought he was fooling them, but they were fooling him. They let him feel successful as a double agent precisely so that they could get the needed information out of him.

This is my experience of mental filters. No matter how hard I try to be in control of my reactions, I find myself playing out the motives of underlying wounds. I may even get underneath one layer of interpretation and say, "Ah, I see! I was being selfish when I took all the attention from my colleague. This time I will make sure to give them the attention." Then, after doing that a few times, I may realize, "Wait a minute! Now I am trying to give them the spotlight so that I can be seen as a martyr. I am still being selfish!" When we free ourselves from one filter, we may find another filter waiting underneath. We still find ourselves trying to bring our lives under control. Our every effort seems to be a new attempt at feeling less vulnerable.

I want to be a good person, but even my attempts to be less selfish or angry can somehow lead me back to selfishness or anger. Like Spyder, this psychological maze appears to be incredibly crafty. It has infiltrated all aspects of our being and is constantly trying to undermine us. Much of the time we are being played, and we don't even know it.

Finding and Choosing Happiness

So how do we get to happiness? Regardless of one's state of wealth or privilege, our filters make happiness elusive. Although there is a basic minimum level of wealth that correlates to greater happiness, beyond that level happiness appears to be a conscious choice.

I live a life of substantial privilege in an area of the world with a high standard of living. Yet most of my life is spent in some form of struggle. Even amidst the benefits of modern plumbing and sanitation, fresh vegetables from the farmers market, two cars in the driveway, and enough bedrooms in my house for everyone in my family, it's easy to feel grumpy and disappointed with life on a regular basis.

I find it fascinating how much suffering in my life comes during moments when I actually have everything I need. Where does the unhappiness come from? The average human in the developed world today has vastly greater luxury than our ancestors only a couple hundred years ago, yet happiness seems somehow just out of reach.

Some years ago I took a trip to Hawaii with my parents and my family. If I can't be a peaceful person when on vacation in Hawaii, then how can I expect to be peaceful anywhere? Yet on this trip, instead of gratitude and joy I became caught in my reactions to everyday things.

When we arrived on the first day, we had to divvy up the rooms. We were blessed to have rented a home on the water with multiple bedrooms. My mom said to me, "I don't have a preference where I sleep. You and your sister can pick the rooms." But I didn't believe her. I was concerned she would be resentful of me if she didn't get what she wanted. I asked, with an edge to my voice, "But what do *you* need?" I'm not proud of the selfishness of my response to her. I was nervous about not getting what I wanted and nervous that she would be upset. I was not at peace in myself.

My mom didn't react too strongly to my mood, but my sister pulled me aside afterward and said, "I think you could have been more gentle with Mom." I had come off as both insensitive and selfish. Why couldn't I just feel happy that we were in Hawaii? Why was I so reactive? I think I was sensitive to my mom's moods and was trying to preemptively avoid the conflict I saw coming. Yet I had just created the conflict. I had not been vulnerable. I had thought I was breaking old habits but was instead playing right into them. Spyder was one step ahead of me.

It can be hard to enjoy even the most beautiful surroundings if we are a mess inside. Paradise is not a place on the outside. Paradise is what we create inside when we become free of our filters and reactions to the situations happening around us.

Fast-forwarding to this year, Dana, Ellie, and I returned to Hawaii, this time with Dana's family for her dad's eightieth birthday celebration. I was determined not to repeat my previous experience in paradise.

I know that I can sometimes be triggered by my in-laws. That's to be expected, right? But I knew it would be even harder for Dana to stay calm with her parents, because she had so many more years of history with them. On our way to the airport I said to Dana, "Whatever happens, I don't want to add to your stress." I wanted to avoid being reactive.

The trip began with ease and I was filled with gratitude to be in Hawaii again. But on our third day my patience was tested. I had the idea of an all-day excursion to Waimea Canyon, the "Grand Canyon of the Pacific." But it was more important to me to have a pleasant time together as a family. So I enjoyed the morning reading peacefully outside on the lanai, unattached and satisfied. We had decided we would go to the beach as soon as my baby niece woke up from her nap. But in the late morning I suddenly overheard hints that the grandparents had changed their minds.

This caught my attention. I had accepted that we weren't going to Waimea Canyon, but I at least wanted to go to the beach. I had prioritized being together as a family and waited around patiently, but now I felt frustrated. If we weren't waiting for them, what were we waiting for?!

My filters were activated and my brain automatically switched into action mode. I was frustrated, resentful, and worried about missing out. This was that same selfishness creeping in. I thought to myself, "Maybe some of us could still take the drive to Waimea Canyon, but time is running out. We need to get on the road as soon as possible!" I put my book away and ran around the house preparing for the excursion. I didn't notice that I was quickly getting attached to my new plans. Spyder was at work.

The family was now making plans to go out to lunch, but I said to Dana that I wanted to go straight to Waimea Canyon. It *felt good* to be pushy with her because I was angry. Yet Dana was now put in a position of choosing sides. Her parents, or her husband?

My pushiness backfired on me. In response to my rush, Dana's dad slipped on his shoes and said, "OK, I'll meet everybody in the car." But I wasn't ready yet! I was trying to plan for our road trip, so I ended up rushing out the door with my daypack hanging open and my shoes untied. When I got out to the car, I lost my cool in front of Dana and Ellie. "I hate being left behind!"

Ah, so that was what was going on. This right here was the key moment, a significant branching point on the tree of possibilities holding the potential for transformation.

I didn't feel like I belonged in the group. All morning I had been uncertain whether I should do my own thing or go with the herd. I was flashing back on my own family upbringing with four other siblings, and I remembered feeling left behind a lot—whether I was last to the dinner table and missed the best pieces of pizza, or last getting my skis on and almost didn't make it on the ski lift. This experience was reminding me of frustrating and lonely experiences I had when I was young.

I could see my behavior having an effect on my family. Dana seemed to be withdrawing into herself as she tried to navigate her stubborn dad and her upset husband. Ellie seemed to be shrinking into silence to avoid receiving my unjust wrath. She didn't want to pick sides between Dad and Grandpa. These dynamics were at their crux. The trail of time could be bent toward healing, or it could be bent toward perpetuating the wounds.

When we are at our weakest, when we are hurt and frustrated, we are right where we need to be on the path of transformation. This is when it is most important to look inside and understand ourselves. These may be moments when we feel least worthy of love. We may feel like we've let everybody down. But these moments are the most precious of all. Most of daily life allows us to avoid doing the work of healing, change, and growth. But in these moments of weakness and pain we have the chance to break the habits of the past and forge a new future.

I could feel that this pattern stretched beyond this moment. How I chose to navigate the conflict could have repercussions for years or even generations to come. It didn't actually matter what we did with our day (the content), only how it felt as we did it (the context). If I chose to feel (misdirected) anger, my daughter would learn how to "manage" me, how to survive within a toxic emotional environment.

I didn't want to create this pattern for Ellie. I too would suffer from this choice. My anger kept me from experiencing intimacy. It kept me from receiving the compassionate understanding of the people who love me.

I had an exciting chance to dissolve this pattern. If I could find my way through feeling-space from anger and loneliness to vulnerability and self-compassion, I would model a healthy way of managing my emotions. This in turn might create a different pattern for Ellie. Maybe she would be vulnerable and compassionate with her own children, and an intergenerational chain of suffering could be weakened.

The wounds of the world are at once immensely universal and yet right here in front of us, ready to be changed. In this simple situation lay an undercurrent of discontent that lives in the very culture itself. It is right there for us to deal with, and yet how do we change it?

As I sat in the car, I stopped talking for a moment and listened to what I was feeling inside.* I felt many urges. I wanted to complain again about Dana's father. I wanted my wife and daughter to know how frustrated I was and how I had been wronged. I also wanted to apologize for losing my cool. I wanted to hold my daughter and say, "Don't be afraid of me! This is not my fault!" But it *was* my fault. It was my temper that had erupted. The only thing I could do was ... stop.

I went through the ARGH! process. I had to accept that my day had been derailed, accept that I lost my cool, and accept that I had already caused some damage to my closest relationships. Then I was able to recognize the pattern. I realized that behind everything I wanted to say was an emotion. There was hurt, and anger, and fear, and resentment. But underneath the emotions I caught a glimpse of something more important. How did I want to feel at the *end* of the day? While Waimea Canyon was a cool idea, what I really wanted was to feel connected to my family and joyous about our vacation. If I could release my goal of getting to Waimea Canyon, I had a chance of changing my experience. I had already made the situation difficult, and it was clear that I wouldn't have the day I expected to have. Instead, I focused on growth. How could I learn about my own patterns of isolation and feeling like I don't belong? How could I help make the situation a positive one for Ellie and Dana in the long run? I had to let go of my attachment to my plans. I also had to let go of feeling mistreated. If I held onto resentment toward Dana's dad, I would not be able to find joy.

Letting go of resentment, hurt, or anger is a simple leap—you just stop feeling the feeling—but it is not easy. I was embarrassed about overreacting, and it's just easier to remain grumpy. But staying grumpy would be trying to hide my weakness from Dana. She already understood everything that was going on. She wasn't judging me, she just wanted to get back to having a good day herself. Although my feelings fought against it, my logic told me that she would be immensely relieved to just forgive the whole thing and get on with it.

---

* Incidentally I also used the LORRAX process from *Living in Flow*—Listen, Open, Reflect, Release, Act, (X) Don't give up/Repeat the process.

In that moment, I found the strength to simply choose happiness. I didn't bury my pain. Rather, I expanded the space inside me to be able to contain my pain without being taken over by it. I thought about how I wanted to feel and forged a path to that feeling. I became grateful for the sunny day and our special time together.

Happiness is not something we create, it is something we *find*. We find it by fumbling around in our inner emotional space. We can think of each emotion as a different room in feeling-space. As we navigate our feelings we move from room to room. To get from the room of resentment to the room of happiness, we first leave the room we are in and then explore the other rooms.

Sitting in the car with Dana and Ellie, I searched my body. I felt shame and anger at being left behind. I felt resentment for sacrificing what I wanted to do for people who didn't even care. But those weren't the feelings I was searching for.

I felt the warm, humid air on my bare arms. I saw the beautiful, moist clouds passing overhead. I remembered when Ellie had told me earlier that morning how I was her "favorite parent" because I went swimming in the pool with her every day. This led me to the room of sweetness and appreciation. From here I was able to find the room of happiness.

We don't conjure happiness. We wander around the rooms of our house until we remember the way back to happiness. The rooms of resentment or jealousy still exist. We don't have to demolish them in order to find happiness. All the rooms are there, and we choose the rooms in which we spend our time.

During lunch the group decided to drive to a nearby waterfall for the afternoon. Going there would be fun! I had found my happy place, and I was ready to talk about having fun together. I didn't simply say to myself, "I am going to be positive." I had found genuine positivity by healing the underlying negative feelings.

Shifting out of negativity is not about something we *do* but about something we *feel*. It's about having enough space for both positivity and negativity, and being able to choose which feeling we want to express at a given moment.

In the car ride to the waterfall, we listened to an audiobook of *Charlotte's Web*, the classic children's book by E. B. White. Charlotte is a barnyard spider who saves the runt pig Wilbur from slaughter by weaving "miraculous" messages like "Some Pig" and "Radiant" into her web. Sometimes finding happiness means listening to your inner Charlotte instead of the sabotaging messages of Spyder. We don't deserve happiness, but we don't *not* deserve happiness. We simply choose to find it.

Exercise: Choosing/Not-Choosing

To find the "happy rooms" inside of us, we can trust our inner sense of what we want. It may seem selfish, but by staying aware of what we want in each situation, we can find a flow-like way of being that is neither selfish nor unselfish.

This can be summarized in a few simple steps that we'll call the Choosing/Not-Choosing exercise. These steps can be practiced when you experience disruption. When an unexpected situation shows up, notice what you are inclined to choose, and the feeling it brings. Notice what you're not inclined to choose, and what feeling it would bring if you did choose it. Then reconsider your choice and go for the feeling you want. Be honest with yourself. Repeat the process frequently, so you stay connected to your inner sense. Here we'll frame the disruptive event as an invitation to step outside our box.

Invitation/Disruption

What am I choosing? Feeling? Why?

What am I not-choosing? How would that feel?

Reconsider?

What is the invitation or disruption?

The first step is to identify when an invitation happens. By *invitation* I mean the opportunities that life sends you to do something different than you are already doing. This may not be easy to notice, as we usually focus on what we are already doing and exclude information that doesn't fit.

If I am in the middle of working and my daughter wants me to see her artwork, this is an invitation. If my family gets on a spontaneous video call with relatives, this is in invitation. If a colleague suggests a new idea for my project, this may be an invitation. If someone suggests that we carpool to work, this is an invitation. If we are encouraged by a friend to switch our career, this is an invitation. We may already have a plan, but the circumstances of life are suggesting something different.

What am I choosing? What's my motivation? How do I feel?

Once you've noticed that an invitation has occurred, ask yourself objectively, "What am I choosing?" Then try to notice why you chose that, and how it feels.

An example might help illustrate this. Our shelter-in-place experience during the COVID-19 pandemic inspired Dana, like many people worldwide, to learn how to bake bread. Yum! When she pulled her very first loaf out of the oven, she called out, "Do you want to come cut it with me?" I was busy on something, so I said, "No, that's cool," and stayed at my desk. But then I asked myself, "What am I choosing, and why?" I was choosing to stay focused on work at my desk, because I was worried that I wouldn't finish my work in time. How did the choice make me feel? I was a little bit lonely sitting at my desk, and disappointed to miss out on the fun.

What am I not-choosing? How would it feel?

Next ask "What am I not-choosing?" How it would feel to choose that instead?

When Dana's bread came out of the oven, I was not-choosing the "breaking of bread" party in the kitchen. But now that I check in with how I feel, I see that it would be fun to be there. I would feel like I was an integral part of my family.

What do I really want?
Now ask "What do I really want?"

If I am honest with myself, I want to be there with them. I want to mark this special moment together and feel like I am part of the group.

Reconsider my action
Finally, I reconsider my next action. Is there something I'd like to do differently?

I decided to make a quick note of where I was in my work and head into the kitchen to join the fun. I felt happy and excited. It was a simple, momentary reprieve from the chronic worry about my work. And wow, was it delicious!

This has become a tradition for us. My life is enriched every time we celebrate the sacredness of the moment that the bread emerges from the oven. Now I always stop what I am doing to come into the kitchen. It brings me a smile and recharges me.

This can happen at work too. On a video project, my collaborator Alex asked me to find stock footage we could use that would lessen his time spent creating content from scratch. This was an invitation to see the project in a new way. But I didn't heed his advice, because I felt worried about the cost. The choice to ignore his suggestion made me feel a little more stressed, because there was more work on our plates. I was not-choosing to follow Alex's advice and buy the needed footage. How would it feel if I changed my mind? I would feel more excited and relaxed, feeling like we were in it together. So what did I really want? I wanted to trust Alex's input and find the easiest way to the final product.

Unfortunately our momentum petered out and Alex got busy with other projects. When I asked him for feedback, he said he would continue to be excited about the project if we could do as he originally suggested. It wasn't too late for me to reconsider my choice. I made a budget to purchase the footage, and I felt a renewed team energy for completing the project.

These questions help me recognize what I really want. They help me become more authentic with myself and others and come at the world from a happy place. But if I am constantly going for what I want, isn't that reckless and irresponsible? Isn't that selfish? Although at first my motivation may be selfish, if I keep asking myself these questions I find it helps me approach life from a wholesome, connected, integrated place.

For instance, let's say I am helping my friend pack for moving, but I don't really have time to help. I'm choosing to spend my day at my friend's house helping them pack. I am doing it because I feel obligated and want them to know I care about them, but I feel resigned and stressed by my own obligations.

I am not-choosing taking care of my own business first. It would be a relief to change my mind. What I really want is to take care of my own obligations first and see what energy I have leftover afterward. With this in mind, I call them up and let them know I have to change the agreement. I feel worried that they'll be upset with me, but I also feel more motivated to tackle the challenges of my day because I haven't overcommitted.

This may at first seem selfish. But a while later, after having accomplished a couple things from my own task list, I run through the questions again. I'm choosing to stay at home and take care of tasks, because it eases my worry about my life. It actually feels a bit lonely and boring. What am I not-choosing, and how would it feel to reconsider? I am not-choosing time spent with my friend. Maybe it would feel refreshing to change my mind and be able to help them a bit. Now it feels good to take a break from my work and help my friend. I call my friend and let them know I'll come over to help once I get to a stopping point in my work.

There is an inner room where we find joy when helping our friends. But if we're not there yet, where are we? If we are anxious about something else, that anxiety is blocking our ability to make joyful choices. It's about timing. It's not that we should simply follow our whims. But why not try to find the inner room where you want to reside? If you are not in that room yet, make the decisions you need to make to get there.

All the things can be accomplished in a natural order. We can trust our desires, for if we stay honest and keep up a spirit of inquiry, all the important things will find their place.

Love and Give

When I wake up in the morning, before I think a single thought, I scan my memory for dreams. I want to remember the adventures I had been so totally immersed in only moments before. When I have good dreams, I start my day refreshed and confident. When I can't remember them, I am more uncertain, and when they are negative, I start my day with a pit in my stomach.

Once I have remembered what I can from my dreams, my real life comes flooding back to me. I remember what's going on in my life and, most significantly, my feelings return from their overnight hiatus. I might feel nervous, or excited, or disappointed. Wherever I left off yesterday is the starting point for today.

Sometimes something good happened the day before, and a feeling of hope and positivity floods me. A smile appears on my face. Most days, though, there is more ambiguity to what I feel. There is a hollowness in my belly, a raw aching that is more than just the need for breakfast.

I ache for clarity. I ache for the confidence to know I am doing the right things with my life. I ache from the uncertainty that haunts me and the wish that I could peer through the veil and know what is really going on in life. I ache for satisfaction and completion, which are ephemeral feelings at best. My stomach's ache tells me that I am still alive, still vulnerable, not yet complete in my mission. It tells me that there is work yet to be done.

I naturally try to soothe that raw aching. I find my mind searching for evidence that I am OK. When I review my email inbox, it is with a hope that something in there will truly make me feel good. When I receive email from a fan, it eases the discomfort in my belly a little. If

I receive an email with good news, I feel euphoria, a hit of feel-good chemicals in my brain. If I receive a bad email, the pit in my stomach solidifies.

I read my email seeking to be propped up by what I read. I am looking for people and events in life to make me feel good and stroke my ego because I am insecure. I don't really know what I am doing, and I want reassurance that I am OK, worthy of love, that my accomplishments are enough. The pit in my stomach reminds me that my future is insecure.

Each day I wake up on this Earth insecure and vulnerable. Will I remain safe? Will I make a blunder and lose my money or my friends? Will my body heal from disease or unease? Will my mind recover from its frustrations? Will the people I love be OK? There is no escape from the feeling of uncertainty in my stomach.

The fact that my sense of uncertainty and vulnerability eases when I get a complimentary email or personal comment from someone tells me something important. Why am I seeking reinforcement? What underlying message am I trying to fend off in my mind? Rather than being compelled to seek messages that soothe my insecurity, is it possible to become more open to that feeling and live without the reassurance?

This life we all have is a life of service. Although I was taught this at a very young age, I only recently really began to understand how it felt to live this way. When I was a kid, my parents followed the spiritual teacher Swami Satchidananda, from whom I learned the tradition of yoga. Yoga is largely known in the West as a physical exercise routine, its popularity due in no small part to Satchidananda's influence. But yoga in its entirety is a collection of mental, physical, and emotional practices that train us to become more aware of ourselves. It is a science of mind, body, and spirit.

Satchidananda's yoga was a life of service. "Love and give, love and give," he said. I could not understand this. How could this be enough? Did this mean I would live a more meritorious life if I were to spend my time engaged in volunteer activities? Should I give more time to charity? These principles didn't really resonate for me. I mean, I have very deep respect for those who give of themselves selflessly in this way. If you are one of those people, thank you for your great contribution to bettering our community. It's just not my way. Then how could I love? How could I give?

But after the experience with the Amazon fires and persevering through my own self-critical walls, I understood better what Satchidananda's teaching meant, at least for me. Everybody wakes up every morning with some version of that pit in their stomach. My wife and my daughter feel that raw vulnerability; my parents and friends do too. My connections on social media, the people who need my help, and the people who are helping me all face the same experience.

Everybody is seeking to feel better, everyone is struggling with their own inner critique that makes it hard to get up each day. To love and give means to keep going even when I feel bad because others need me. To love and give means to nourish the part of others that needs tenderness and confidence. To love and give means to help alleviate the uncertainty that they feel, allowing them to be more confident and effective themselves. To love and give means being OK enough that I can help somebody else feel OK. It is a tool that helps me get out of my own misery and back into flow.

Now I find Satchidananda's message deeply meaningful. It means I accept insecurity in my own life. I don't expect the world to fully affirm my worth. I will always wake up with a pit in my stomach. The world is a mirror, and I contain mysteries. Certain people and things appear synchronistically in my life to show me what I am thinking and feeling, to show me my hidden self. As long as there exist within me storms of criticism and judgment, that will be my experience.

I understand Satchidananda's teaching as "giving" nourishment to others. And to do this I have to tame my inner critic, which is the inner source of my insecurity. Is this what the "love" part is about? I said that I will always experience the gnawing of uncertainty in my stomach, but maybe this is not true. What I mean is that no amount of evidence from the world can remove that discomfort.

But maybe the deepest experience of love and self-compassion can vanquish this insecurity. In my experience, the feeling of love doesn't come from arguing the counterpoint to the critic. Love doesn't seem to take the checklist of our faults and monitor our progress against it. Love breaks the rules and runs naked through the field.

When we love ourselves, we do not say to ourselves, "I know I failed my test, but I will make up for it." Rather, we say, "I know I failed my test, and that hurts." Love seems satisfied to feel the emotion. From this simple acceptance, we might then be able to say, "I accept myself and my present situation, and I look forward to the next chance to learn."

Our filters tell us we are OK only if we improve, but without those filters we experience the rawness of who we are. Is love this unbroken expanse of wholeness inside our chest?

I find myself frustrated by the casualness with which the word *love* is used. For me, love is best captured in other, more descriptive phrases and feelings. When I use the word *love* I am not referring to a thought or an emotion but an experience to be had. We do not *feel* love. We *experience* it. It is raw and plain, a direct experience of our circumstances.

Experiencing love doesn't mean being nice, or being helpful, or being supportive, or listening to others. I can experience love when I am telling my daughter she can't have dessert, or firing an employee, or telling my friend I can't help them move, or ignoring someone I care about.

The experience of love is, although vulnerable, also immensely powerful. We do not rely on others for validation of the experience of love. As a parent, for example, I feel free to love my child to the point of embarrassment whether she permits me to or not. The depth of my love is shameless. I don't need my daughter's approval to love her. I am going to do it whether it annoys her or not. I feel more confidence as a parent than perhaps anywhere else in my life, not because I am necessarily good at it, but because I experience so much love while acting as a parent. The experience is not contingent upon anything else.

When I wake up in the morning with a sense of emptiness at the core of me, this too can be thought of not as a feeling but as an *experience*. There are many different feelings I may have—insecurity, worry, frustration, inadequacy—but all these constitute an experience of gut-wrenching emptiness.

No amount of evidence will counter the experience. It is like trying to mend a broken heart by eating cake. Information and proofs cannot alter the fundamental experience I am having.

There is a leap that must be made. We can continually gather evidence to counter our negative thoughts about ourselves, but at some point we must leap, without validation, into a different experience.

How do we make this leap? We experience love. We experience our complete selves, without holding anything back. We experience wholeness. Love is not a feeling to be sought after—love is an experience to be practiced.

As we gain more power to choose and direct our inner resources, what comes next? In the following chapter we will delve into the difficult work such as grieving and facing pain that can be accomplished when we have the presence of mind, heart, and being to do so.

10

THERE *IS* A PURPOSE

When I was in sixth grade, the librarian took a few of us under his wing. Mr. Conway was a tall, wiry man. I don't remember being particularly close to him, but we were a small community, and he lived right behind our house. I am sure he knew me much better than I knew him.

The year was 1985, and Mr. Conway had managed to purchase three personal computers. They were set up in a small wing of the library, and I was invited to join him once a week for a computer class. The computers operated on DOS, the rudimentary text-based interface used before the more modern desktop design came out. Computer class consisted of playing around with a program called Logo. You would program a list of geometrical commands like "move forward 100 pixels, turn right 20 degrees, move forward 50 pixels …" and then watch the "turtle" draw your commands. We could use loops to make it repeat our commands any number of times. Due to the loops, the pictures were always very symmetric. Patterns were repeated over and over.

I found the class somewhat boring, yet it had a dramatic influence on my thinking and life path. The patterns we created might look like little snowflakes, or trees, or crystals. This was my first exposure to fractals. A *fractal* is a mathematical structure that repeats a set of basic rules over and over again, often showing unpredictable structures and patterns in the overall picture. The pattern you input at one scale—the basic "forward, turn" commands—can show up as a pattern at a larger or smaller scale. They are incredibly beautiful and can be used to generate realistic scenes of nature in computer animation.

In college I became more deeply interested in fractals and the idea that large-scale structures reflect patterns at smaller scales. This is true in the way patterns of galaxies at the highest levels reflect patterns of neurons in the brain and arteries in the skin. I began to wonder if this can be true of human society too. Is it possible that the intractable problems happening at a societal level have their seeds in patterns at the individual level? This idea has become only more firmly implanted in my mind over the intervening years. If this is the case, maybe it can provide some new ideas on how to bring about change. Maybe the patterns that show up in society can be used to show us our own hidden wounds, places where we can heal and grow. Thus, maybe there is a purpose to the mystifying misdirections that happen in life. Maybe there is a pattern to our wounds.

It all starts with pain.

Our Personal Pain Becomes Global

During the same time that I was taking computer classes with Mr. Conway, my family changed dramatically. As I have mentioned, I lived with four siblings at my mother's house. My stepdad had two sons, each just a few years older than me, who had lived with their mom since I was very little. It was at this time that they came back into my life permanently. I had been the oldest kid out of three in the household, and now I was the middle kid out of five.

This was a very impactful period of my life. My brothers and I are very close now, but it wasn't smooth at first. I had difficulty adjusting to having new people around. My brothers weren't always kind to me, and I am sure I wasn't always kind to them. My parents' attention was divided now between the five of us. I found I had to fight for myself, defend myself, grab for things before there was none left. I was blessed with a richer family in the long run, but what I chose to do at the time was build walls. I became less kind, less open, less vulnerable.

We become fractured by life. Not only as kids but also as adults we make unconscious choices when we feel threatened by daily experiences. These choices can persistently change the way we think about ourselves and others. We might choose to hide our culpability for a problem at work, or choose to get angry at a parent because they say something we don't like.

Life comes at us fast, and most of us move through small traumas every day without stopping to process them. Our personality can easily ossify around our unconscious ways of managing these experiences. We methodically cover up our mistakes at work, or we stop listening the minute our father opens his mouth. This just becomes "the way things are."

Our personal fractures are also reflected in our communities. Our brittle economic and social structures seem to reflect the fears that many of us carry in our hearts. If a person is consumed with the question "Is there enough pizza for me?" as I was, it propagates outward. Modern economics is about divvying up resources—material resources, land resources, or even attention resources—amongst individuals. The attitude of "there's not enough for me" and the pattern of divvying things up live inside of us, starting so far back in history that it is difficult now to envision another way of thinking.

Threading the Needle to a Better Life

Story contributed by Jenna

I live with epilepsy, and I was working in a job where the stress made my epileptic symptoms much worse. Yet the job provided health insurance, which was crucial for me. I got fired from the job, but I went on to use my insurance to have brain surgery, which cured my seizures and transformed my life. It helped me refocus on my goals and complete my internship hours to get my license as a therapist. This allowed me to move back to the small town I had dreamed of returning to. My quality of life has improved many-fold through these changes.

The question "Can I trust you?" is also reflected in our society. Even though I fought with my brothers, I learned that they had my back too. When we got in trouble with our parents, they didn't try to pawn off blame on me. Other relationships I learned not to trust, such as Kaylen at school who called me a weenie. These patterns show up in my life today. They influence how well I collaborate with peers and with my managers at work. It may influence who I elect to government or who I hire to run a department.

Our fractures are fractals, patterns that appear at every level of existence. They start at the most intimate level as ruptures in who we think we are, and then appear in our relationships and even beyond.

Our filters *seem like* reality, so much so that we have trouble believing that someone else could feel differently than we imagine. When we act according to our filters, we entrench our perceptions and build our lives on those foundations.

What about when our filters influence our politics? Most of us vote for people we agree with. It seems like a no-brainer, right? I don't think so. If my filters are limiting the information I base my opinions on, what good is it to elect someone who has the same blind spots as me? Our personal filters get magnified by the leaders we choose.

Does it seem ludicrous to vote for someone who doesn't agree with us on policy? The point is not that we shouldn't fight for our beliefs. Rather, it is that our filters are often distorted by pain. It can be hard to see how our choice will really work out.

Peace begins at home. In the years following my struggles in Los Angeles, Dana and I rebuilt our lives together back in the San Francisco Bay Area, where we had originated, and eventually got married. Early in our marriage, I think Dana and I both felt like we had to defend ourselves to each other. I had to make sure to protect my spare time and my career, and I think Dana had her own values she felt she needed to protect. The first five years of our marriage were difficult. Gradually I learned to trust her more. Through experimentation, honesty, vulnerability, and risk, I found that when I took her needs into consideration, my own needs were taken care of too. The more we each took the stance of wanting the other to get what they want, the less we needed to defend our position.

When situations triggering doubt come up, it is an opportunity to examine myself inside and wonder what I am afraid of. For instance, one summer I was very excited about being offered a slot to play at a music festival a couple months away, whose date hadn't been decided yet. When Dana's family invited us to go on vacation over the likely festival weekend, I wanted to say no because it might conflict with the festival.

But we committed to the vacation. I had come to see that she always stood up for me as best she could. It felt better to trust in flow than to be held hostage by the fear that I wouldn't get what I need. Sure enough, as the date of the vacation approached, the music festival ended up being rescheduled for a different weekend anyway. How was I to know this would happen? I couldn't. This is the power of synchronicity in strengthening our trust in relationships. We can expect synchronicity. This allows us to drop the fear and focus on working together.

Advocating for our opponent seems ludicrous in the public domain. If I don't defend my proposals, my opponent will replace them with their own. If I advocate for my opponent's point, how can I be sure they won't capitalize on the advantage? Of course we can't be sure. Relationships are about trust. We grow trust in each other over time, through being trustworthy. Having a political system that is polarized and devoid of trust is a result of small, repeated untrustworthy actions over time.

This is the recursion of my Logo game. If one's basic program or way of thinking has the subroutine "Do what it takes to win" somewhere within it, then in a thousand little ways trust will be eroded over time. We shouldn't be shocked when we see this play out. It is just a reflection of the programs running inside each of us. It is a sign of the times.

I don't think we will be able to recognize the right paths forward until we have a frame of mind ready to see them. It is important to see our own faults and admit our mistakes, accept that we don't have the answer, trust each other with our vulnerability, and be the first one to let down our guard. In becoming willing to peel away our filters and let go of the things we so surely believe, we become able to trust each other more. Whether it's climate change, COVID-19, or systemic racism, transforming the way we think can make a meaningful difference.

Pain is expressed differently at every level. Our personal pain of having been unheard, unsafe, or treated unfairly can propagate upward to create a collective environment that reflects our personal pain. The paths forward lie within us. Recognizing the impact of personal filters on our collective politics is something we all have a part in, and it leads to ways of thinking and choosing that have the potential to increase our capacity for collaboration.

The Healing Power of Being Alone

Learning about how our personal filters influence our choices makes it easier to be connected instead of isolated. But I don't want to create the impression that isolation is a bad thing. We don't need to constantly strive for more connection. Intentional isolation and quietude are not things to run from. They are fulfilling and important for personal development.

Being alone with ourselves can be a very uncomfortable experience. My discomfort usually comes not from other people but from the conversations I have in my own mind. When I am alone, these difficult conversations are still there. Conversations with myself become more pronounced and clear when nobody else is around. I realize I am angry or scared, defensive or resentful. I do not need someone else around to have these emotions; they happen all by themselves inside me. I might receive information from the news on my phone, or remember a recent conversation I've had with a friend, or find a dirty bowl left on the kitchen counter; there are so many ways to trigger the conversations I have with myself. Even in our quietest moments, our mind is bringing images and memories to our attention.

The responses that internal conversations trigger help us understand ourselves. They show us how we have learned to interpret the world, the layers of personality that hold us back from feeling fully alive. Since nobody else is around, we can be assured that anything we are feeling is ours to feel.

Fulfilling a Dream of Getting Pregnant

Story contributed by Iris

My husband and I decided we wanted to have a baby, and after a few months it became clear that we needed help from a fertility doctor. For my job as a government lawyer I was expected to spend a year working on assignment, but for fertility treatments I had to see my doctor frequently, so I applied to work in a nearby township and was accepted. It was all arranged when my boss, without explanation, reassigned me to a township ninety minutes away. The doctor's office closed at 6 pm, and it was impossible to continue fertility treatments. We grieved the lost opportunity to get pregnant, and eventually decided not to pursue it further even though the work assignment was temporary. The stress and grief had been taking a toll on me. Nonetheless, I became pregnant without medical intervention three months after starting my new assignment. The spontaneous job reassignment ended up allowing us a less stressful pregnancy experience than we had feared.

As I write this morning, the sun has not yet risen. Usually this is a time of wonderful quiet and concentration for me. But today I find myself struggling for ideas, my mind constantly pulled away. Where is it going? The COVID-19 pandemic has changed the mental environment. People are online at all hours of the day. Online conversation is ongoing and constant; information is moving manically. I find myself wanting to be a part of that, not wanting to miss out. I think of this person I should respond to, or this link I found that I should share, or this other idea I want to write about on social media. There is no quiet, no downtime in my mind. People have begun to isolate themselves intentionally with social distancing, and suddenly there is a tremendous increase in online activity. As external commerce shut down, the interior chatter of a billion connected minds became cacophonous.

This is what it is like to practice meditation.

When we put away our list of chores and cease our busyness, we expect things to be boring and quiet. Instead, as things get quieter outside, we usually experience a rush of thought and emotion inside. These are the contents of our minds that were already there, the low-volume signals that were previously drowned out by the external noise. The processes that govern society reflect the processes that govern each of us alone.

Being alone is necessary. I find it hard to track what is going on inside my own mind when I am actively engaged in a conversation. My filters respond too quickly to be noticed. In taking space from the conversation, the conversation might replay itself inside my mind and I find clarity on what I was feeling. This is a helpful way to notice what filters I am hearing through.

It is easy to feel trapped by thoughts and feelings, stuck in the maze. We might respond the same habitual way every time we speak to our parents, for example. Alternating between time in conversation with those people and time alone can help us catch a glimpse of the filters. Once we see them, we can use the processes in the previous chapters, such as Synchronicity as a Process of Healing, the ARGH! Process, or Choosing/Not-Choosing, to use our circumstances to break our habitual cycles. In aloneness we have the quiet space for the roar of our mental conversation to become noticeable.

Aloneness can be beneficial for identifying our filters or developing mindfulness. Alone time can be time spent in traditional meditation, but it can also be time spent reading quietly, or walking out of doors, or exercising at the gym, or snuggling with your children, or talking deeply with a romantic partner, or sleeping. These all can provide you moments of clear perspective on your filters so that you can gradually unwind them. What is crucial is that the activity allows you to be aware of your emotions as they happen.

We should not simply fill in the loneliness with connection, as if trying to fill a hole. True connection emerges out of honoring our loneliness. Within the quiet lie our true emotions, and if we aren't familiar with ourselves in this way, how can we live authentically? Our hope for true connection is in developing a clear inner channel to our loneliness. True isolation can be very healing, yet it takes more courage as well. Healing our filters means being quiet enough to see our patterns of absorbed trauma and being compassionate and honest with ourselves. We may have to re-feel the feelings we are avoiding—shame, betrayal, frustration—in order to stop holding them inside.

Being alone with ourselves takes tremendous courage, but it is worthwhile. Once we can be alone with ourselves we are more capable of bringing harmony into our communities.

Healing Is Living with Grief

I vividly recall where I was in April 2010 when the Deepwater Horizon oil rig disaster occurred in the Caribbean Sea. I was working as a programmer, and the atmosphere in the office was quiet and focused. I sat at my console watching live footage of the broken drill at the bottom of the ocean, spewing high pressure oil onto the ocean floor. The scene was appalling, impossible to believe really. When I learned that it would take three months to fix it, I was in shock. You mean I could sit here with three months' worth of coffee and just watch oil erupting into the bottom of the ocean all day, every day, for ninety days? How is that possible?

This seemed like a devastating blow for life on Earth. Would our oceans ever be the same again? But over the coming days I noticed that others in the office weren't sharing in my shock. Business was progressing as usual, with just an occasional mention of the disaster in passing as we would chitchat about the news. I felt alienated. I felt out of place. I could barely work. The world was falling apart under me, and I couldn't find a foundation. I needed to share this experience with my team, but there was no way to do this.

I had this experience of shock again in March 2011 when an earthquake and tsunami destroyed a nuclear power plant in Japan and radiation poured into the ocean for weeks. I had it again when fires devastated Northern California, Australia, and the Amazon rainforest in 2017 and 2018. These were all overwhelming experiences. If we are not shocked and grieved, what is going on with us? We are all impacted by these disasters. The only thing I can figure is that we must have trouble seeing that impact in the moment.

Why can't we grieve together? What stops us from feeling the traumatic events that strike us globally? We might be worried that "we can't just express grief all the time, we would never get anything done, society would not function!" I understand that concern and suggest only that there must be a middle ground. What would a healthy relationship to grief look like?

Collectively we seem to have pushed grief into a corner so it won't distract us. Maybe we thought we could eradicate it altogether, like smallpox or polio. But grief has followed us through all those years, waiting until we were quiet enough to hear its still, small voice. It provides us with a persistent opportunity to grow. "Hear me," it says, "I am not afraid of my sadness. I am grateful for what I have gained and lost, and I don't want to run from myself any longer." Through the muffled barrier of our filters—*I need to meet the next deadline, I need to reach the next milestone, I need to please my parents or my friends*—we hear that gentle plea. Don't be afraid of me. I am your hurt. I am your original sadness. I seem threatening to you, but I will free you from the chase.

Feeling grief doesn't look the same for each person. We often think about grief as the uncontrolled expression of emotion, like me sobbing at school after my cat died. If grief takes over a work situation, that's probably too much. But we can live with grief without disrupting the fabric of our lives. Grief is not just one but a collection of feelings. Grief is about having access to all the emotions that arise without blocking any of them. At work I simply wanted to know that others were having a hard time. I wanted to share in my struggle. We can acknowledge these difficult feelings without collapsing underneath their weight.

Living in flow is living with grief. When loss arrives, we remain open to everything that happens to us, ready to respond with clarity and compassion. Living with grief is not depressing. Grief is what connects us to the fact that we are alive. It is the reminder that our time here is precious. Grief unlocks our passion for living and galvanizes our commitment to change. Grief can cut through our filters and help us choose wisely. Our sorrow, when we allow it to be felt, has the power to change our behavior.

Climate change itself is an externalization of the attempt to avoid grief. If we didn't feel insecure about our financial future, might it be easier to change our fossil fuel economy? If we weren't afraid of losing our standard of living, might it be easier to accept short-term inconvenience for the sake of solutions that last? If we weren't afraid of having difficult conversations about equity and fairness, might it be easier to understand how economic inequality exacerbates the climate crisis?

These fears are strategies for deflecting grief. Climate change is not just a collection of physical problems. It reflects our inner state. It is an expression of our limited ability to handle grief.

Charles Eisenstein says,

> In order to reverse the course of ecocide, we may have to consciously choose a healing path. We cannot count on collapse to compel us. In order to choose it, we need to change the conditions from which we are choosing. To change those conditions, we need to implement a different economic system and understanding of nature, and more importantly, we need to recover our empathic ability to feel. Therefore, the issue of environmental degradation and climate change cannot be separated from the need for social, economic, and personal healing.

> To recover our ability to feel is going to hurt, because so much pain is out there waiting for us to feel it. It has been sequestered away, suppressed within ourselves and kept out of sight globally. On the outside, walls of cement and razor wire, walls of disinformation, walls of prisons, walls of gated communities, walls of historical blindness, and walls of complicit silence keep the dominant culture unconscious of the suffering of damaged peoples (human and otherwise). On the inside, it is false hopes, delusions, addictions, and pharmaceutical mind-control agents.[1]

I was unable to find connection at work, so I shut down. But there is another way. We can expand our filters and incorporate grief into our daily lives. It will inform our choices and make us more compassionate human beings.

Opening to Grief

I often find deeper insights into my behavior lying in the little moments. A single incident in an otherwise uneventful day can be an invitation for exploration into the depths of who I am. Healing can come anytime and from anywhere.

I had recently partnered with a volunteer, Tan, to help me on a project with a tight deadline. He had just gotten a promotion at work, so I knew he was busy, and I was grateful for his help. But this relationship was to trigger a place of vulnerable insecurity for me.

I woke up one morning as the deadline approached still having heard nothing further from him. I was feeling increasingly anxious about whether we would finish the project on time. I checked email repeatedly all morning and kept composing—but not sending—emails that I hoped didn't sound impatient or ungrateful.

I had a hard time focusing at work. I kept wondering if Tan was mad at me. Had I done something wrong? I realized that the feeling was familiar. I felt the same worry about Tan as I had felt about a partner on a previous project last year. Then I remembered the same experience with my team at my previous corporate job, and about the members of my band in high school, and even about my best friend Charlie in elementary school. Charlie ended up just kind of disappearing from my life, and I never learned why. Apparently he didn't like me as much as I liked him. I was having a grief reaction, feeling abandoned, unreciprocated, lonely.

Rather than obsessing over Tan, I turned off my phone and didn't check my email for the rest of the morning. I lived with the discomfort for a while. Later in the day, as I was listening to my daughter share about a friend of hers, I had an insight into my own experience. I was hurt. Why had Tan not even bothered to follow up? Was it me? What had I done to make him angry? In addition to feeling hurt, I felt angry. I imagined how good it would feel to never call him back. I wanted the satisfaction of being the one who left first.

But I couldn't help myself. I turned my phone on again to check messages, and a notification appeared. Tan had left a message very early in the morning, but it hadn't been delivered until now. It had some-how not arrived when it was sent. Ugh. So I was wrong, Tan *had* called. Flashbacks sped through my brain. I thought about the time wasted on phrasing and rephrasing my email—"I know you're really busy," scratch that, "I haven't heard from you, when is a good time to talk?" scratch that, "How do you feel about our deadline next week, do you think it's possible?" scratch that—and now I realized all of that was unnecessary.

I was confused and frustrated by my emotions. The *fact* was that he *had* called. He still liked me, and he still liked the project. We were simply having communication issues. My previous theories about being abandoned and unreciprocated were full of holes.

Why had his call not come through earlier? This was a meaning-ful coincidence that threatened to undermine our relationship. It was a chance to choose between the familiar path of resentment and self-pity, like I had with Charlie, or something new. Here was a situation in which there really was no problem. It was up to me to choose my filter.

Resentment is a familiar blanket. If you try, you can probably feel it right now. Give it a shot—picture someone who you feel has treated you unjustly, and you will probably feel resentment for them. Though it feels like *your* resentment, it is not personal to you. It is a tool you have inside you. It is a set of chemicals that you allow into your brain and bloodstream. It is a pattern you picked up from your elders. It is a coat that you can pick up to keep yourself warm whenever you want to feel its cheap comfort. You can even think of it as a parasite that feeds off your hurt feelings.

Grief can help in these moments. The desire to punish others is a strategy to avoid grief. When hurt and sadness are knocking at the door, how do we let them in? It can help to acknowledge how much time we have spent feeling resentful. That was wasted time. Holding onto resentment is like throwing good money after bad. If we continue to be resentful, we just waste more of our lives.

Admitting how much of our lives has been wasted in negative pat-terns allows us to grieve. To see the pain we have chosen to hold on to, that is grief. Lost time. Lost energy. Lost days and even years of life. To see the people we have alienated, and the pain we have caused others. That is grief. To see the failed plans we gave up on. That is grief. To comprehend the magnitude of wasted effort in life. That is grief.

It requires great strength and courage to tolerate all that grief. Yet we have all had this experience. This is why living in flow is living with grief. The constant acceptance of loss gives us the potential for freedom in the future.

Thus, I had a rotten choice between resentment and grief. No wonder I have so often chosen to be resentful in the past. Choosing grief feels terrible! Here I could stay in resentment a while longer, making Tan pay for making me feel bad. But if I do this, I hurt him and repeat my familiar ways of failing. Nothing would change.

Can I open the door when grief is knocking? Sometimes I can. Often I cannot. So I try to walk softly. I don't want to damage my relationships through blindness. When someone hurts me, I try to remind myself that I don't know what is going on for them. I try to remember to approach people with curiosity rather than forming judgments or assumptions.

Unexpressed grief can lead us to do things we regret. How are we to know what we are really feeling when our filters can be so deceiving? Our grief is a treasure. It is our most precious clue to understand who we really are. The closer we keep it to our heart, the more confident we can be that it won't catch us by surprise and cause us more grief.

Content versus Context

Everything that happens can serve a purpose. Pain shows us where we are wounded. By pulling away from the content—a person's disapproval of us, or the possibility of getting sick during a pandemic—and seeing what underlying pain is being revealed, we have the opportunity to heal. We survive in the moment by focusing on the content—dealing with the logistics of the crisis. But we level-up our skills to our next level of functioning by understanding the context—understanding how our own filters have exacerbated the problem. Our suffering can serve us for good. Our struggles can be part of a satisfying journey of growth.

Serving Heroes on the Front Lines

Story contributed by Matt Upton

I purchased a travel trailer intending to spend time camping with my grandkids over the summer. Ten days later the COVID-19 pandemic changed everything. All my consulting contracts abruptly ended. I pondered how I could continue to serve people in need, and an idea dropped into my soul: I can go where they are and serve alongside them. I was deeply moved by the efforts of the "Lunch Girls and Guys" around my state who serve meals to students, so I hooked up my travel trailer and went to support them on duty. They had gone from obscurity to becoming essential workers on the front lines. I have now pulled and lived in my trailer for 77 days, served at 40 different curbsides in 36 different school districts, including on a school bus, in the rain, sleet, snow flurries, and blistering heat, for a total of 4,400 miles. I've assisted these amazing workers in providing 317,775 meals to students across the state.

When pain occurs, say divorce in a family, it may feel like an unfortunate and hurtful experience, but it can also help us embrace important changes in our lives. Even when the content is devastating, the context can be meaningful and may lead to a greater sense of who we are. At each new level, we open ourselves to new gifts and greater experiences that were not possible when our pain was in the way. By seeing the context of our problems, we more easily find purpose in them. When we understand our context—who we are and what we care about—we can handle the content of life better. Our filters become easier to navigate. We may be suddenly able to see why we were doing things a certain way, why we were clinging to a particular viewpoint. Then we become free of our old patterns of thinking.

If our possible futures are laid out on a tree, our filters lead us one way or another through its branches. Choices made in resentment or anger lead in a different direction than choices made in compassion and for-giveness. Our default reactions are usually the ones that no longer serve us. My default reaction is often resentment or self-pity, another person's default might be to be overly compassionate, yet another's might be to shrink down and become quiet, while a fourth person might always let others go first. Our ability to take something *other than* the default path is a test of strength. It may take effort to restrain our anger or fear. It may take effort to stand up for ourselves when challenged. It may take effort to let someone else get their way without any promises in return. These are the choices toward peace, the roads less taken.

In *Living in Flow* I called these *bold actions*. The more emotional effort needed to take an action, the more dramatically we can change our life through that choice. It takes emotional strength and willpower to look past the content and respond to the context. Doing so allows us to see what is really needed to defuse tension and resolve a situation in a way of maximum benefit for everyone involved. When we are not caught up in our default reactions, we become more peaceful. This is how we shift away from conflict and toward a sense of companionship and trust.

The problems of the world seem big, but it doesn't take a whole lot of inner changes to create significant outer change. Once we truly under-stand our own behaviors, it become easier to make different choices. When we *feel* the change at a personal level, it sticks at a global level. In the next chapter we'll examine what it looks and feels like to have successful experiences of growth and change.

11

WHAT IS THE LEAP?

On New Year's Day of 2020 my family was looking for something to make the day special. Dana had brought up walking across the Golden Gate Bridge, something we had never done together. We were taking care of our baby niece that day, so we invited my sister and her toddler, but she replied that they wanted to do something less crowded and simpler. She invited us to go walking in a local park, which was my daughter's favorite option, but I wanted to make it a special day and walk the bridge.

The first hint of trouble occurred on the freeway as we approached the bridge. The line of cars getting off the freeway was backed up a quarter mile and moving like a glacier! My inner weather erupted in a crash of thunder as I realized that this may have been a big mistake. Pema Chodron describes what I felt: "When really strong emotion comes up, all the doctrines and beliefs that we've held on to seem kind of pitiful by comparison, because emotions are so much more powerful. So what began as an enormous open space becomes a forest fire, a world war, a volcano erupting, a tidal wave."[1]

But Dana was cool and collected, so I followed her lead and checked my attitude. In my old thinking, I would blame myself for getting into this mess. I had tried to impose my will and missed the flow. My choice seemed like a bad one, so I was frustrated.

But I had another option. I imagined the whole tree of possibilities. Somewhere on that tree were branches on which we had a wonderful day. How could I get to those? Who would I have to be right now to carve away at the situation until we found a "good day" buried within it? Chodron says, "Instead of struggling to regain our concept of who we are, we can touch into that mind of simply not knowing, which is basic wisdom mind."[2]

Twenty minutes later we made it to the front of the line of cars, only to realize there was a detour directing us away from the bridge. There was no way to get there! This meant more stormy weather inside me. I felt the urge to lash out in frustration, but I still wanted to find a path to a good day.

This is when I practiced the ARGH! process. I accepted the reality of being trapped in a nasty traffic jam in the middle of nowhere. I accepted that it was my fault. What was the pattern? I had been *absolutely certain* that this was the right thing to do, even to the exclusion of other advice. I could see how this certainty was a familiar feeling. What would it mean to grow here? Instead of jumping to solutions and seeking more certainty, maybe I could try listening better to others? As it turns out, the traffic patterns slowly funneled us away from the bridge and toward an alternate destination, the Marin Headlands. I knew there was a beach there with a beautiful view of the water. Everyone in the car was OK with that, but a minute later I found myself in stopped traffic again. This time it was easier to avoid getting sucked into my inner stormy weather. I wanted to doubt our choice and just turn around and go home—there was that certainty again—but Dana was calm, so I remained calm.

Twenty minutes later we were taking a beautiful stroll up onto the bluff, soaking up the rejuvenating views of the Pacific Ocean and the wild wind of the western coast. We had found our good day, and I had healed a stressful pattern that had an impact far beyond that moment.

There are two ways of creating: one is to build something out of nothing, and the other is to chisel away from what is already there, to reveal the possibilities within it. Through our choices we navigate all the possibilities, maybe to a dead end or maybe to a beach. When we try to create the life we want, it takes force and control. It's exhausting, and I don't know about you, but I usually create problems for myself. But if we go for the feeling we want to experience, and see ourselves chiseling away from all the possibilities, we experience more ease and ultimately more satisfaction. How do we see the latent possibilities? We can repeatedly ask ourselves, "Am I listening to all the information? Am I stubbornly holding on to an idea? What filters are controlling me right now?"

The more we open to the wholeness of who we are, the less our filters limit who we can become.

Wholeness Is an Activity

Twenty-something Stefani Germanotta was performing at a nightclub in New York City. She was a seasoned burlesque performer, having worked with mentors and honed her craft for many years. She had developed a formidable stage presence and nurtured a sense of confidence with her skills. But those things alone wouldn't get her where she wanted to go. She had to make a leap.

On this night the audience was particularly difficult. There was a group of college fraternity kids being disruptive as she tried to settle into her groove. She sat at the piano but couldn't begin until she had captured the attention of the room. Trained in the dramatic flair of burlesque and wearing "an amazing outfit," she followed her gut in a pivotal moment of inspiration. She began to take off her outfit. The room got quieter. When she was finally sitting at the piano in her underwear, she had the audience's attention.

In that moment she discovered who she wanted to be as a performer. It was not premeditated—and I think this is key—it was a leap that felt appropriate in the moment. She was connected to her inner experience, and she *felt through* her choices in the moment. This was not simply the discovery that "sex sells." This was Germanotta's personal integration of her mental and emotional training. She was making a choice on the outside that felt right on the inside. She describes it in the following way: "It was a performance art moment, there and then. You see, you can write about it now and it will sound ridiculous. But the truth is unless you were there in the audience in that very spontaneous moment, it doesn't mean anything ... in the context of that moment, in that neighborhood, in front of that audience, I was doing something radical."[3]

She became known as Lady Gaga. In that moment, she felt the pull we all feel at certain pivotal moments, an urge to do something important but which feels risky and terrifying. If she had planned out her striptease, that might have been a different experience. Who knows if it would have worked the same way. The spontaneity of the experience seems to have allowed this to be an experience of healing and integration. Eliciting shock is not something Lady Gaga does to manipulate others into applauding her. It is an expression of the nakedness of her soul.

The leap to wholeness is about the experience itself, unscripted and unanalyzed. It is the application of what we know, a shift in personality that can't be understood mentally. It must be felt. This same immersion into experience is needed for athletes when they make the leap into flow, or politicians when they stand up to represent others. It is felt by skilled teachers when they engage their students' hearts and minds in the learning process, or by sincere lovers who peel away layers of guardedness with each other.

The leap to wholeness is a moment of healing.

I'd like to reframe the experience of healing. I'd like to encourage you not to think about healing simply as overcoming injury. Don't silo emotional or mental healing as unpleasant. Healing is the underlying thread tying together our daily experiences. Healing allows us to find joy in relationships and helps us achieve our dreams. Healing is the activity of becoming whole, and being whole is to be engaged in your life in every possible way. When we heal, we get to experience new things. We get to experience new feelings of connection or excitement because we are able or willing to do something we weren't before.

Lady Gaga acknowledges in an interview, "I still sometimes feel like a loser kid in high school and I just have to pick myself up and tell myself that I'm a superstar every morning so that I can get through this day and be for my fans what they need for me to be ... I feel like any other insecure 24-year-old girl. Then I say, 'Bitch, you're Lady Gaga, you get up and walk the walk today.'"[4]

Healing is a disconcerting mix of the unpleasant and the pleasant. Together they form a whole. When traumatic experiences hold us back by threatening our perceived self-worth, healing is a process of becoming clear on our actual worth as a human being and living it through our actions. The difficult emotional weightlifting and mental gymnastics that are often associated with grief and trauma are tools that can be employed in the process of healing. But they are not the point. If you could jump right to wholeness, the painful and difficult work would be unnecessary.

The hard work of becoming whole is shedding all the things that are in the way. The distance is not far, but the leap can be tremendous. Lady Gaga's years of training in burlesque theater prepared her for her moment of transformation. In the end she was not so different than who she was at the beginning. It was there inside her all along. That final stage of healing, specifically the mental or emotional kind, is a leap into something unknown and scary.

This is the leap to wholeness. We do not have to seek permission of our friends or family, we do not have to achieve certain credentials, we do not need to earn it or deserve it. The wholeness I am talking about is there when we are ready to take it, unconditional, powerful, and vulnerable.

The Moment When We Choose

Wholeness is, in theory, just a breath away from us at any moment. Yet my journey toward wholeness has taken a long time, and continues today. Like Dorothy's ability to return from Oz at any time by simply clicking her heels together and saying "There's no place like home ..." I probably could have found my way to greater wholeness and happiness long ago if I really understood the way. But finding my way to wholeness has meant understanding my path and the choices that got me here.

I would say I had a great childhood, and I am grateful for it. Yet buried within the experiences of childhood lie the roots of our patterns. How did we react to pain and disappointment? How did we navigate things that were not fair? No matter how positive our childhood, it can be valuable to understand how the details of those years influenced the filters we carry as adults.

In some ways, I cut myself off from others at an early age. In my family I had no clear partner, no best friend. My parents had divorced when I was an infant, and I spent time between both families. In one household, I was one of five kids. The older ones I was most close in age to, but they were older boys from my stepfather's first marriage, and they were bonded together in a way that didn't include me. In the other household, my sister was ten years younger than me. In the critical years of elementary and middle school, I saw myself as a loner.

I remember being at my dad's house, when my younger sister was still very little, creating fantasy games outdoors in order to keep myself company. I had only a few peers living in my neighborhood, and I wasn't particularly close to any of them.

Because I spent half my time at each parent's house, I didn't feel like I belonged at either house. I became very independent. I became afraid to ask for things that I needed, like money to take the bus. I preferred to find my own way of getting around. I rode my bike, walked, or hitchhiked. I made choices to be lonely. I learned to cover up my vulnerability and pretended like I had it all together. Since my teachers or the principal never called home with problems, I guess there was no reason for my parents to be worried about me. But I wish I could do those years over. I would try to let people in more frequently, not keep myself distant and my needs hidden from the people who cared about me. I would allow myself to feel like an integral part of my family and community.

As an adult it is not too late to change. The work of healing can always be done. I had an opportunity to heal this for myself on the trip to Hawaii with Dana's family. We had had a beautiful day at the beach, where my daughter was learning to surf. As we watched her navigate the waves, I had stuck my wallet carefully into one of the zippered pouches of our backpack. Later that night at dinner I realized I didn't have my wallet. I looked in the backpack and it wasn't there. I began to scour the house to try to find it. I even considered making a night run back to the beach to look for it.

Things began to get worse for me. I started to feel a bellyache coming on. Dinner had disagreed with me, and I felt increasingly uncomfortable as I went inch by inch through the rental car and through our luggage, trying to retrace my steps to find the wallet. Then I walked back into the living room and, demoralized, collapsed into a chair. Dana was telling her family about my lost wallet, and they were empathizing with me.

A Twisting Life Path

Story contributed by Rev. Dr. Raymont Anderson

I moved to another state to be with the person I was considering marrying. I was very distraught and torn because leaving Pittsburgh meant I could no longer be the primary caregiver for my father. After a long conversation with him and receiving his full blessing, I moved. Almost immediately the relationship started turning sour. After a few months of growing more and more distant, and a life-threatening experience with a former student of mine, things came to a head and I moved out onto the street. All of this struggle led me to enroll in the MFA Theatre Pedagogy program at a local university, reviving a dream I had relinquished years prior! I started teaching public speaking, then got offered to be the public speaking coordinator for our department, then started an American Sign Language performance company. At graduation I was offered a job with a Deaf performance company in Washington DC, where I moved and eventually was led to becoming an ordained minister within my spiritual community! Through many losses, I was directed to finding my Purpose as a Spiritual Teacher.

"He put it in a zipped pouch, so it's hard to imagine it fell out at the beach," Dana said.

"Did you look in the fold-up chairs in the back of the car?" Dana's sister asked.

It was here that I noticed my habitual reaction. Rather than being grateful for their help, I picked up a book and began to read with an air of nonchalance. Letting them see that I was worried would be vulnerable. Would they criticize me for being careless and forgetful? I stayed silent.

I didn't actually notice all these thoughts in the moment. But I did notice myself thinking, "I don't need their help." I noticed that I felt ashamed. This wasn't the first time this had happened. Some years earlier we had been traveling together to Mexico and I had realized on our way to the airport that my passport was expired. The whole family had witnessed me struggle with that uncomfortable dilemma, scrambling at the last minute to get my passport renewed. Now here they were again coming up with creative solutions for how I could replace my lost driver's license before flying home at the end of the trip. It was so embarrassing. I just wanted to handle it on my own to avoid the shame.

But this time I didn't. This was a moment of choice, and I decided to let them in. I didn't mention the wallet but instead said, "My belly is hurting." This one little phrase went against decades of habit. I felt like an embarrassed little kid with a bellyache, unable to care for himself.

To my surprise, I received sympathy from everyone in the room ("Oh, I'm so sorry to hear that"). They had noticed that I was cranky, but now it was more understandable. I didn't realize how much people had learned to tiptoe around me when I isolated myself, but now I had let down my guard and they were right there to keep me company. Everyone really wanted to help me feel better. I didn't feel alone anymore.

It dawned on me that I had never really allowed that experience as a kid. I had learned to take care of things for myself and was often unable to feel the caring that people had for me. Love has two forms. In its potential form, love can exist under the surface, but unless the conditions are made for it, it doesn't flow out and express. If I had remained grouchy about the wallet and the bellyache, my family's love for me would not be less, but it would have found no way to express itself. Their love might have turned into resentment at my emotional distance. But love can also take an expressed form. In order for love to be shared and felt together, we must express it in some way. In order to allow love to be expressed, we have to be willing to receive it, and for this to happen it helps to be vulnerable.

The paradox in becoming more vulnerable with my family is that I become stronger and more resilient in my life. By opening myself up to my family, I also get out of my own self-pity. Perhaps there is symbolism in losing my wallet that night, as if it was a chance to try on a new identity. Having let them into my problems, I felt more motivated to solve the problem. I went back upstairs to use the bathroom, where I noticed my bathing suit drying in the shower. I picked it up to feel the pockets for my wallet, and suddenly had a visceral flash in my body. I remembered placing my wallet in my damp bathing suit pocket earlier when we had arrived at the car.

This triggered an insight. I had brought not one but *two* bathing suits; where was the other one? I had left it outside to dry. I went down to the patio lawn chairs and fumbled in the dark. My bathing suit had fallen behind the patio chair, out of sight. On the ground under the chair in the dark I felt the stiff edges of my wallet!

I went back into the living room and triumphantly tossed my wallet into Dana's lap, saying, "I found it!" Everybody erupted in laughter and relief. They were happy for me. By letting them in on my problems, I had made space for the celebration at the end. It occurred to me that by living much of my life without sharing the struggle, I had also missed out on sharing victories. My experiences of success had become drier and drier as I simply moved on to the next thing, without experiencing real satisfaction. Now I was getting the thrill of shared accomplishment. Others knew what I was going through, and they got to share in the joy and relief that comes at the end of the journey. In sharing that experience, I got to experience it more deeply as well.

Life is like a game that must be played but cannot be won. Finding my wallet feels like winning, yet it just allows me to return to the other problems of everyday life. We never transcend the problems, or win the game once and for all. Instead, we can ask ourselves whether we are playing it wholeheartedly. Just finding the wallet is one thing, but sharing in the experience with my family is a whole different experience. It is a new level in feeling-space.

Sometimes we "find the wallet" and sometimes we don't. The content of the goal changes each time. But along the way we can examine whether the choices we make are helping us to grow. Playing the game is about each choice. Is our choice leading toward greater wholeness? Is our choice steering us away from conditioned childhood patterns and toward an adult capacity to love and be loved? It is not the finding of the wallet but the sharing of the struggle—the context—that brings satisfaction.

As a child, I was confronted with the feeling of being left out, and I chose my response. Through reacting to the weather of the moment, I created a climate of isolation that lasted into my adulthood. Yet deep change can be a simple shift at the ground level that has broad repercussions for quality of life. It was a simple four-word phrase, "My belly is hurting," that unwound that pattern. The right choice at the right moment has the power to shift the climate even after years of repeated patterns.

Losing my wallet and having a bellyache can be seen as synchronicities. They may be commonplace events, yet for me, at that time, they were the key ingredients to a significant shift in the quality of my experience. Whether consciously or unconsciously, I was ready to break free of my feelings of isolation, and these experiences helped me do that. The events themselves were not unlikely, but their timing was perfect. When I was ready for the leap to wholeness, the right experiences showed up to get me there.

Healing as a Way of Living

Our wounds are like chasms in our souls, and practices of self-awareness can help us scope out the best places to cross. But no matter how we do it, whether by building a bridge or with a running jump or dangling from a zip line, we must still cross the chasm. Because of the inner preparation we have done, the old fears may no longer consume us. Yet they are still there, just as fear was there for Lady Gaga sitting alone at the piano with a rude audience.

We can face our fear by reprogramming our thinking so we are not seduced by our filters. In my experience fear doesn't really go away, but my mind stops jumping to self-protection. Because I have more self-compassion and less self-criticism, there is less risk involved. I am not really threatened. I am just trying to level-up myself to a more authentic and satisfying way of living. Gaga's preparation allowed her to avoid being overwhelmed by self-criticism, so she was able to find flow. The fear was still there, but her mental control allowed her to cross the chasm.

The chasms that divide us from ourselves can show up in many forms, and I wonder if these are the same chasms that divide us from each other. This is why healing is so important. As a child, my sense of wholeness was altered by bullying. I learned to back down from confrontation out of fear of physical pain or shame. Even today I find myself making mistakes in business because I am trying to avoid backlash from a client.

This fear of bullying has never gone away, but I have become more aware of what I am afraid of and have reduced the power that those filters have over me. When a client or friend pushes me around, I have more choices. I don't feel trapped. I can choose differently than I would have ten or twenty years ago. All the mental preparation in the world can't take the leap for me. When the confrontation occurs, it is the actual *being different* that matters. Having built the bridge or set up the zip line, I must actually hoist myself over the gap and make the crossing.

The experience of finally arriving into a healed state is a leap to wholeness. It is a sudden surge—a "quantized" shift, in the language of quantum physics—in which we become the person we want to be, not because we earn the right, but just because we want to. We do not wait for evidence to build. We change on the inside first and begin to see the evidence build as a result.

So what does the process of healing look like? I think all our life experiences occur in service of the growth of our soul. We are thrashing our way through feeling-space ungracefully, groping blindly in the dark with our hands outstretched, trying to understand why life happens the way it does. Through healing we gain greater confidence, deeper compassion, stronger connection, more security, easier forgiveness.

When we become more confident, we become more powerful. Becoming clearer on who we want to be with our parents, we might find more ease with saying the necessary things in the necessary manner to improve the relationship. Becoming more confident at work, we might more easily find our voice when we have something to contribute while being self-assured enough to leave space for others to find their voices too. In either of these cases, developing compassion-based confidence helps us navigate to more complex, inclusive, and adapted decisions that benefit more people. Healing is a crucial part of having the power to create the change we wish to see in the world.

When we develop deeper compassion, we are able to see the suffering we cause for others through our careless choices. We have more space to "be," instead of acting from a compulsion to fix the world. When we feel compassion, we and everyone around us can relax a little. We are less likely to criticize unfairly or direct our inner conflict outward onto other people. Starting with compassion for ourselves, our gentleness flows outward to affect everybody in our lives.

Through establishing stronger connections, we have more peaceful lives. Each filter removed is one less barrier between our eyes and the eyes of those we speak to. Each conversation becomes an opportunity to feel more deeply, and the more frequently we experience genuine feelings in the company of others, the less isolated we feel. Connection is a mutual experience. When we really see someone else, we also feel really seen by them.

Through healing we feel more secure in our experiences. Uncertainty nags at us, challenging our faith and stimulating our feelings of isolation. But security grows out of connection. When we feel connected to other people, we feel more confident in the face of the unknown. We even feel more confident with life itself. When we feel connected to an inner source of guidance—for me, this is the process of meaningful history selection that reflects my own inner purpose—we can act with more confidence even when the circumstances are shifting under our feet. We don't have to feel less secure just because we are less sure. We will never live in a more certain world, but the quality of our experience of uncertainty changes as we become fully expressed in the world.

And through healing we find it easier to forgive. After the long journey, when we are more confident in ourselves, more compassionate with ourselves, more connected to people and to life, and more secure in our path, it becomes natural to forgive. As we gradually shed the filters of interpretation in our mind, we also shed the burdens of resentment that we carry, and guilt for our own mistakes. Forgiveness allows our future to be free from the past. Forgiveness allows wholeness to emerge from a previously fractured life. Charles Eisenstein reminds us, "That moment of humble, powerless unknowing, where the sadness of an ongoing loss washes through us and we cannot escape into facile solutioneering, is a powerful and necessary moment. It has the power to reach into us deeply enough to wipe away frozen ways of seeing and ingrained patterns of response. It gives us fresh eyes, and it loosens the tentacles of fear that hold us in normality."[5]

Healing is not a luxury, it is a way of life. Healing is what we are all doing, all the time, taking us to new levels of experience. By making the leap to wholeness in ourselves, we can be a part of the healing of our community and our planet. In the following chapter, we'll expand our intention beyond personal healing to include our families and our communities.

12

WHOLENESS
IN COMMUNITY

Synchronicity and Privilege

On a Sunday afternoon I left my house for Oakland, to attend a service for the Thrive community, a New Age spiritual group focused on social justice and racial equity. I had been in a rush leaving the house, and planned to come right home afterward, so when I couldn't find my wallet I just trusted that I would be fine. I had no fear about not having my wallet with me—a subtle example of my privileged thinking. I grabbed $20 from my desk drawer and left. Shortly before the entrance ramp to the freeway, just a couple minutes from home, I noticed blue and red police lights in my rearview mirror.

I felt a rush of adrenaline, a familiar feeling anytime I see a police car behind me. I remember when I was sixteen and had my first car. I really struggled to afford the car insurance every six months. Then I got my first speeding ticket and was overwhelmed by the expense! Now, thirty years later, I felt the same queasy disembodiment when I saw those lights in my mirror, not because I am afraid of the police but afraid of the cost. I didn't have my driver's license on me, and I had broken the law by using my phone while driving, yet my mind was focused on the *cost*. That is a sign of privilege.

A Black person may have different concerns than I do when stopped by the police. For me, it didn't cross my mind that getting pulled over and not having a license would even cost me a ticket. In the worst case, I would have to mail in my proof of license later. For a Black person, driving without a license could turn into being arrested or losing their life.

I have written a lot about the positive effect that synchronicity can have on our lives. Even though it is a neutral phenomenon, not good or bad in itself, when we become aware of its effects we can navigate life more smoothly.

But I struggle to understand the role privilege plays. Is a world of synchronicity associated with a privileged worldview? What sort of meaningfulness can it provide for someone who faces systemic racism, or xenophobia, or poverty, or homelessness?

I wish I understood this better. My sense of empowerment in the world has come out of an experience of privilege. I take the view that synchronicities occur in our lives as reflections of our own choices, yet in my life the playing field has been fairly tame. It is easier to learn from life if we are given leeway to make mistakes. Even the big misstep that I described in the first chapter, when I ended up in the hospital, did not undermine my potential for a good future. I had the support of well-off people within my community who provided a safety net. It is easier to experiment and learn from our experiences when there is leeway to make mistakes.

In leaving the house without my wallet, I was trying to make sure I wasn't late for an event. That is a privilege. What if I was Black and that was a job interview? If my life depended on having my wallet with me, then I might have to be late for the interview while I search for it. Thus, the privilege of making mistakes allows me a better chance to get ahead. I have a greater chance of being there when synchronicity happens.

What about the privilege of non-urgency? Isn't it easier to see synchronicity in daily life when we know we have time to try things? As a white person, I experience a flexibility of circumstances that may not be experienced by others of different race or socioeconomic group. I have rarely lived month to month. Usually I have time to think ahead and plan the next steps of my life. Even the times in my life when I was in need, I had a network of people and support (such as friendly banks and friendly landlords) whom I implicitly knew I could trust to help me. Without ever experiencing the fear of being on the street next month, I have been able to invest in education and opportunity. Synchronicity has shown up consistently along that path.

Would that be true for someone of lesser socioeconomic opportunity? How is synchronicity relevant for someone who lives with urgency forced upon them every day by the circumstances of life? I don't know. I hope you, my readers, may have something to contribute on this question. I believe that people of all backgrounds need the hope and wholeness that synchronicity brings. Wherever we are in our struggles, the right circumstances can make a difference. Meaningful coincidences can help.

But can they? If a person is living within an unjust system, they cannot simply rely on synchronicity to keep them safe. They must fight actively against the specific structures causing the injustice. When the real estate industry consistently favors white home buyers, synchronicity alone can't change the policies that make it difficult for people of color.

Racism is not a coincidence. Racism comes from prejudice, and prejudice comes from filters. In *White Fragility* Robin DiAngelo explains,

> Prejudice consists of thoughts and feelings, including stereotypes, attitudes, and generalizations that are based on little or no experience and then are projected onto everyone from that group. All humans have prejudice; we cannot avoid it. If I am aware that a social group exists, I will obtain information about that group from the society around me. This information helps me make sense of the group from my cultural framework.[1]

She adds,

> When a racial group's collective prejudice is backed by the power of legal authority and institutional control, it is transformed into racism, a far-reaching system that functions independently from the intentions or self-images of individual actors.[2]

Experiences that increase our prejudice reinforce filters we have inherited from others. The process of navigating the weather of our experiences, as I wrote about in chapter 4, leads to a limited and ingrained view of life. DiAngelo continues, "From birth, we are conditioned into accepting and not questioning these ideas. Ideology is reinforced across society, for example, in schools and textbooks, political speeches, movies, advertising, holiday celebrations, and words and phrases."[3] Each of these events creates patterns in our thinking that filter out specific *potential* beliefs from the wholeness of who we are.

The beliefs we accept and identify with are incomplete versions of the world. Racism itself is a filter on wholeness. When we look at a picture of a person in the paper, we may make a snap judgment about their personal qualities based upon the clothes they are wearing, the background in the photo, and the look on their face. We are allowing ourselves the luxury of an incomplete picture.

Finding Inspiration for a School Project

Story contributed by Ellie and Sky Nelson-Isaacs

Ellie and I take our usual bike ride along a local bike path, but today we plan to take a detour down to Main Street to buy a new bike helmet. I get my directions wrong and accidentally arrive at Main Street quite a few blocks early. I'm disappointed, since riding among traffic is not fun. But suddenly we pass over some inscriptions in the sidewalk and bring our bikes to a halt. We have just passed a memorial with twenty or so plaques dedicated to Japanese Americans from our local community who were sent to internment camps during World War II. This is amazing, because Ellie is in the middle of a big school project on this topic! We take pictures of all the plaques, and the experience reinvigorates her enthusiasm for the school project.

This, too, is a privilege. To be able to survive and thrive with incomplete information means that the system in which we operate rewards that choice. Journalist Jon Greenberg identifies ten types of white privilege that can be seen in most white Americans' lives.[4] Two of them, "I have the privilege of soaking in media blatantly biased toward my race" and "I have the privilege of being insulated from the daily toll of racism," illustrate this privilege of incomplete information. It is possible—in fact, rewarded—for white people to maintain a falsely filtered view of the world.

Instead, we find that the system favors us, and thus it may seem that synchronicity is at work in our favor. For Dana and me, purchasing a home was a stressful but fun and empowering experience. Little things fell into place just right, like getting unexpected money back from the bank when escrow closed. Synchronicity was on our side. But these experiences all happened within a system built to favor us. According to DiAngelo, "Whiteness rests upon a foundational premise: the definition of whites as the norm or standard for human, and people of color as a deviation from that norm."[5] As white people with good credit due to lifelong access to financial opportunity, we had a team of people making money off helping us. As white people, we were under the assumption that this is simply the way the system works. It is, for us. The gears of synchronicity were lubricated by class status.

There are ways to make a theory of synchronicity and wholeness more relevant to people of different race, different socioeconomic background, or different culture, and this is work that I would really like to see happen. To do so will require input from many people, and I hope we can do this work together.

What We Add to the Stew

When we have an interaction with someone of a different race, it is hard to keep our filters clean. We naturally add our own interpretation to each thing that occurs. What are we adding? We add whatever we are already feeling inside.

If I am angry or fearful at someone due to superficial reasons, I am bringing anger and fear with me into the interaction. If I am rude to someone based on skin color, something is disturbing me inside. This is the power of understanding our filters. Our filters color everything we perceive, and they are directly influenced by how we feel. So what we perceive is as much a function of how we feel as it is a function of what is really there.

I know when Dana has swept the kitchen because I will find a little pile of dirt sitting somewhere on the floor. Apparently she doesn't like the last step of transferring the swept dirt into the garbage can. I take a deep breath. It's a little bit annoying!

Yesterday I left the lid off the coffee maker. Dana was diplomatic with me. "You might want to consider, in general, putting the lid back on the carafe, so gunk doesn't fall in." Between that and the peanut butter jar I left out on the counter last night, I know she is fending off annoyance as well.

On the days when these small annoyances irk me, I know I'm upset about something else. The irritation was already there before they arrived on the scene.

I know this is true with a simple thought experiment. Would I be annoyed at this small issue if I was feeling awe and wonder in my life right now? What if I had just found a new client or gotten a piece of writing published—would I care about the dirt, or would I simply sweep it under the rug? If something small is bothering me, my unhappiness is probably about me, not her.

What ingredients am I adding to the stew?

The same goes for people we see on TV. How many of us sit in front of the television watching Jon Stewart or Tucker Carlson just looking for the next opportunity to be mad? We watch these shows because we have a certain worldview that we think will improve things in the country. But do we really want to create a better world, or are we after the rush of feeling "right"?

Feeling right is a filter too. We do not walk into most situations as impartial bystanders. We are looking for a chance to reaffirm what we believe, and the reinforcement of our filters feels good. Who is able to watch the news these days without a rush of adrenaline? Not me! I seek evidence for my personal theories, information that matches my filters. I want to be able to say, "That person is an idiot!"

But just as with our personal relationships, if our day is going great before this person appears on television then the thing they say will slide right off our back. When people get under our skin and we are looking for a chance to discredit them, it is a sign that we walked into the room unhappy. That was our own contribution to the meal.

The hurt we are trying to heal is not new. It has existed for a long time. We are experiencing emotional pain because of our internal filters. Some of these we inherited from other people, some of them we've generated ourselves. How much of the time that we spend upset is due to baggage from some previous situation? Unacknowledged pain makes us brittle. It is hard to be resilient when we are being pummeled by fearful and unconscious internal messages.

During the COVID-19 pandemic, I felt horrified to see the stock market dive 30 percent simply because people stopped shopping and flying for a week. These are preexisting systemic weaknesses in our economy. Similarly when an offhanded comment from a parent or a slight change of expression in our spouse's face is enough to cause us pain, we know the pain was already there.

When we are happy with life, we are naturally compassionate and resilient. We meet problems with determination and inspiration. But if we won't let ourselves be happy until a certain condition is met, then we bring that unhappiness everywhere. We shall see in the next chapter that in a holistic world patterns become the most important building block. If our life is built on a pattern of being dissatisfied with our job, or resentful of our spouse, or mad at authority, this pattern will appear everywhere. This is the ingredient we add to the communal stew.

If we think, "I will be happy once I get Dad to compliment me," or, "I'll be happy once I am successful in my job," then until these things happen we bring unhappiness with us wherever we go. We walk into the room with some terrible thing on the news and exclaim, "Oh no, not again!!" The filters we carry into the room with us limit our choices. But if we were in an elated mood, we would have access to a much deeper well of possible experiences. We could feel compassion, anger, sadness, and joy all at once without contradiction.

When we are unhappy with ourselves, we cannot find peace with anyone. There is always some condition our happiness depends on. Some item on the list is always missing. It is natural to feel irritation when disruption occurs. But how resilient we are is at least a partial reflection of our level of inner peace.

The False Self

The way I have presented the filters that determine our perceptions and our personality reflects the theory of "the false self" put forward in the 1960s by psychologist Donald Winnicott. The false self emerges when "other people's expectations ... become of overriding importance, overlaying or contradicting the original sense of self, the one connected to the very roots of one's being."[6]

In his book *The Desire for Mutual Recognition*, law professor Peter Gabel describes the false self's impact on modern political culture. Our false self creates an image of who we are in the context of our family, as well as in the wider contexts of our community, our nation, and the world. He points out that a key change takes place in our mind when we shift from picturing ourselves "here" to picturing ourselves "out there."

Gabel says we naturally create an abstract model of ourselves and how we fit with everyone else. But in picturing the world and ourselves "out there," we lose access to compassion and belonging, or what Gabel calls "mutual recognition." To directly acknowledge the existence of another person is to value their feelings as your own. This means feeling the person right there in front of you, rather than projecting a mental image of them "out there."

For instance, if I am in a meeting and someone gives me negative feedback, I may develop a story. "Janet is just so sensitive," or "James can dish it out, but he can't take it." The story casts them in a negative light and reduces the threat they pose to me. Thus, I create an abstract mental model of our relationship that generates a false separation between us.

A similar process generates a false self in relation to our community. Gabel says, "Thus instead of being *here* in direct relation to everyone else here, I imagine myself to be 'out there' with all the others as others, a 'citizen' among other citizens in the imaginary unity of a large group of socially constructed entities."[7]

We create a mental picture that helps us understand where we belong. "But the mental picture of it ... is primarily a kind of externally imposed badge that is a part of the cement sealing me in my separation."[8] Our mental picture of our political environment doesn't include a direct experience of anything. There is no me and there is no you in my mental picture. There is just a layer of filters that helps me think that I understand you, and myself too.

It is hard to look at the world through another's eyes. As liberals, people like me create a persona that we imagine as a "conservative." We think these conservatives in our heads are totally crazy. But then I find myself catching an episode of *The Bachelor*, a show about gratuitous sex that devalues the seriousness of relationships. When I see it I think it's gross, but I just chuckle and turn the channel and move on. I don't feel threatened by it because I understand that only a limited subset of my community really thinks that way.

But I can certainly see how somebody—with values not so different from my own—could see this and be horrified. They would be justified to conclude that "liberal values" are everywhere and they are undermining America.

Ultimately the difference between the political poles in our country may have more to do with context than content, more to do with the filters than what is filtered. This means that if I were raised in a different context, I might feel very differently about the content of the issues. But then, can I really trust my own opinion on the issues?

In imagining social change, we often focus our attention on the issues. But what about the attitude we take in addressing the issues? Who am I becoming through social change? If I am trying to create peace, am I coming from a place of inner peace? Do I feel peaceful? Meaningful social change is more possible when the suggestions being made come not just from a vision of peace in the future but a mindset of peace right now.

What would it take to be at peace in the world and yet committed to change? In my family when one person is grumpy, we all feel it. Everybody starts to feel grumpy. Whatever our initial upset was, it gets amplified by others.

Rather than focus on the issues, we can listen to what people are needing. Dropping our own filters and interpretations, we can hear what is behind the words. I can bring compassion to the context of how someone else is feeling, rather than getting caught in the content of how they are expressing themselves at this particular moment in time.

If we are to be messengers of peace or justice, being peaceful inside is important. We can be fierce and firm in our resolve while remaining calm. This is called *equanimity* in Buddhism.

If I am carrying my own backdrop of unresolved anger, shame, and resentment, I will bring these qualities with me into the stew. When one person speaks on an important issue but injects a little of their own woundedness by saying "Anybody in their right mind would not agree with …," they pollute the dialogue. Gradually the dialogue become filled with garbage ideas and emotions unrelated to the issue.

This leads us back to the filters that guide our perceptions and our actions. Only when our filters are healed and our perceptions are clear can we contribute to the stew without adding toxicity to it. Gabel says, "A transformative politics capable of generating social change requires an emergence into connection with each other, through a coming into mutual recognition, that dissolves the moat that otherwise separates us."[9]

It is through changing our personal interactions that we will change the political crises confronting us. When we anticipate greater connection, we take actions that bring us together and "dissolve the moat." When we watch the news, are we experiencing greater connection? Do we have control over our inner experience so that we are a healing influence on others? If we are caught in anger, ridicule, disbelief, or condemnation, we are being tricked by our filters. Can we step back and be witnesses to the drama but not blinded by it?

In gaining awareness of our filters we can sit with the discomfort of beliefs and values that are contrary to our own. We can anticipate connection even in the face of evidence to the contrary, and we can expect synchronicities to arise that give us the opportunity to come closer together rather than move further apart.

Healing the false self gives us strength and power. The false self is healed by perceiving the wholeness of who we are underneath.

We have spent several chapters investigating holism and filters at an inner level, exploring how we might identify and change difficult patterns. This is part of how I envision the Holographic Paradigm, inspired in the first chapters by the holistic nature of light and the connectedness of time and space. In the next two chapters, we'll return to the science of holism and get into many details of how, in my understanding, the holographic multiverse could work.

PART 3

13

IS PHYSICS HOLISTIC?

We learned in part 1 that certain aspects of the physical world—in both space and time—cannot be reduced to the sum of their parts. This is holism, or wholeness.

Where do we see this in nature? Consider touching the surface of a bowl of water with your finger. Can you do it in a way that does not generate a ripple? No, it cannot be done. The whole surface is affected by even the tiniest interaction.

Given today's focus on reductionism—the view that everything can be understood by studying the parts—you might be surprised to find how many examples of holism there are in physics. Hiding within everyday experiences that you may take for granted are clues to our fundamental connectedness. I find it exciting to look at these examples and reignite the wonder that comes from noticing hidden details of life. Nobel Laureate Richard Feynman helped me see the hidden beauty in plain sight. He said,

> I have a friend who's an artist and has sometimes taken a view which I don't agree with very well. He'll hold up a flower and say "look how beautiful it is," and I'll agree. Then he says "I as an artist can see how beautiful this is but you as a scientist take this all apart and it becomes a dull thing," and I think that he's kind of nutty. First of all, the beauty that he sees is available to other people and to me too, I believe, although I might not be quite as refined aesthetically as he is, I can appreciate the beauty of a flower.

At the same time, I see much more about the flower than he sees. I could imagine the cells in there, the complicated actions inside, which also have a beauty. I mean it's not just beauty at this dimension, at one centimeter; there's also beauty at smaller dimensions, the inner structure, also the processes. The fact that the colors in the flower evolved in order to attract insects to pollinate it is interesting; it means that insects can see the color. It adds a question: does this aesthetic sense also exist in the lower forms? Why is it aesthetic? All kinds of interesting questions which the science knowledge only adds to the excitement, the mystery and the awe of a flower. It only adds. I don't understand how it subtracts.[1]

For Feynman, the relationship between the color of the flower and the insects was a hidden detail. Similarly holism is a hidden detail that we are likely to pass over if we don't know what to look for.*

Let's walk through some examples of holism that you may not have thought of before, and maybe we'll find new wonder in the familiar.

Don't Dawdle, but Don't Rush Either ...

Story contributed by Sky Nelson-Isaacs

I had struggled for over a year with an important part of my research regarding potential energy. I wanted to submit a paper for publication anyway, since I didn't see a resolution in sight. A mentor read the paper and said to hold off on publishing until I could cite more references around a different technical issue regarding our understanding of "time." I was frustrated by the advice, but I heeded it and did some more digging. I soon found an interesting paper that supported what I was trying to say about time. But in reading it I discovered that the author was using a mathematical technique that solved the other problem I had regarding potential energy. Following my mentor's advice solved a problem that he didn't even know I had!

---

* Physicists are aware of holism by two other names: *non-locality* and *duality*. The focus of the work I am presenting here is on duality, which comes from the Fourier transform.

Wholeness in the Propagation of Light

Studying digital media—image editing, video compression, or music creation—is an easy and fun way to get an intuition for holism in the world. To do this, we will first examine one of the most basic phenomena we deal with every day: How does light get from place to place? This is called *image propagation*. Most people take for granted that light can bounce off an object and travel to their eyes, but in fact this can't happen without the timelessness and spacelessness of the frequency domain. This is the field of optics and diffraction theory. Since we're getting technical for a bit, instead of *pattern space* we'll use the correct technical term: the *frequency domain*.

If we have an image—say emitted from a Netflix show we are binging— the basic problem is to describe how that image moves from the space at our TV screen to the space at our eyes.

Movement of light occurs *outside* of space and time, in the frequency domain. To understand this, we have to understand how to break up an image into its parts. When I first learned that any complicated image could be broken down into simpler parts, I was blown away. There is hidden order within even the most diverse forms. We call it *decomposing* an image, because it is much like how a piece of fruit in a compost pile gets decomposed into its basic elements. A piece of fruit looks unique, yet it is made of just a fixed number of basic elements. In an image, the simpler parts are not molecules of sugar and water but "plane waves." Plane waves are just the basic elements of frequency in the frequency domain. Any photograph can in fact be composed by overlaying the right combination of these types of patterns.

Figures 13.1, 13.3, and 13.5 show examples of plane waves oriented in different ways. The *frequency* describes how rapidly the pattern oscillates from black to white and back. Each of these plane waves takes up the whole picture, yet it is captured in its entirety by just a couple of dots in the frequency domain, shown in Figures 13.2, 13.4, and 13.6. This is very economical! *We can describe a pattern that fills all of space with just one tidbit of information.* Placing a dot in the frequency domain tells my computer, "Draw lines throughout the entire picture with a certain spacing and at a certain angle."

FIGURE 13.1. A plane wave pattern of quickly varying white and black lines.

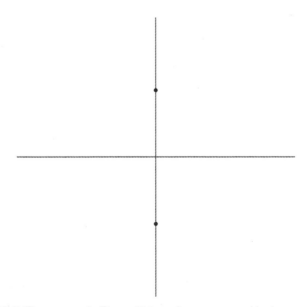

FIGURE 13.2. The pattern in Figure 13.1 can be represented by just two vertical dots in the frequency domain, far from the center. You can ignore the fact that there are two dots rather than just one. This simply indicates that the pattern can be equally drawn top to bottom or vice versa.

FIGURE 13.3. A different pattern with a lower frequency and different direction.

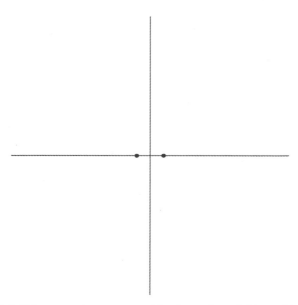

FIGURE 13.4. This pattern is represented by dots oriented side to side, and closer to the center.

FIGURE 13.5. The pattern can exist at an arbitrary angle and frequency.

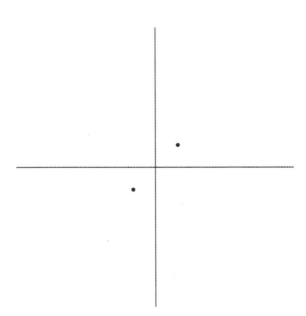

FIGURE 13.6. This pattern is represented by two dots at the appropriate angle and distance from the center.

We can decompose any image, say a photograph of the city in Figure 5.1 into plane waves. Figures 13.7 through 13.11 show partial images made from increasing portions of the total frequencies that make up an image. Encoded within the relationships between these simple patterns is a more complicated pattern that looks like a city. The city image becomes more apparent as we include more frequencies.

FIGURE 13.7. This image contains just a handful of different plane waves, moving in different directions with varying frequency and varying brightness. Only very basic patterns are apparent.

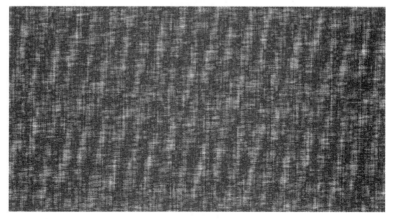

FIGURE 13.8. Now we include more frequencies. Here, too, only very basic patterns appear, although slightly more interesting.

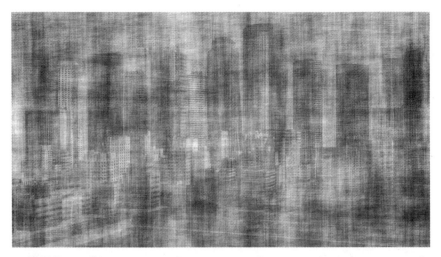

FIGURE 13.9. With even more frequency contributions, we begin to see some of the more complex structure in the relationships between the plane waves, as the shapes of city buildings start to emerge.

FIGURE 13.10. We have now included a majority of the frequencies in the original image. The image of the city is quite apparent, although there are still distortions from the remaining frequencies that are yet to be rendered. The edge of each building is a sharp line whose position is encoded within the relationship between the phases of millions of simple plane waves.

FIGURE 13.11. We now include all 2,073,600 frequencies present in the original photograph. This crisp, distortion-free image is a re-creation of the original image.

Is that not amazing?! An artist can draw a picture of a city using a pen, and in doing so the basic element they use is a pixel of ink. Put the pixels together in the right way, and you see a city. That's the usual reductionist way of viewing the world. Here we have done something very different. The basic element is no longer a pixel but a pattern. Whereas an artist would use 2,073,600 pixels to draw the city, our computer used 2,073,600 patterns. Each of those patterns was drawn over the entire screen. None of them can be associated with a specific place on the image. When you add them all together, they interact with each other, and out of the resulting relationships emerges a hidden order—the city skyline.

You may have heard that when you cut a hologram into pieces, you can still see the whole image in each of the smaller pieces of the hologram. We have just learned why. *Every frequency component is drawn over every part of the film, so all the information is embedded everywhere.*

This is the central way that wholeness enters into physics. The frequency domain is related *as-a-whole* to regular space. One can divide up physical space—such as by cutting the holographic film in half, or by considering you and me as separate—but this doesn't divide up the frequency domain. *In the frequency domain, you and I are both part of a single, irreducible description of the cosmos.*

So we've now described how the image on your screen can be described by patterns. Let's return to the question of how this image gets from the television to our eyes. In the classic textbook on image processing, the author summarizes, "Our analysis of (image propagation) has yielded a space-invariant form of the (image)."[2] The frequency domain that encodes these wave patterns is called "space-invariant." I have called it *spaceless*. It doesn't capture the positions of the buildings but the *relationships between the buildings*. In Figures 13.7 through 13.11, the position of a given building is determined by the relationship of all the line patterns drawn on the page.

How does this image appear when it reaches our eyes? You would probably imagine it moving through the space between the television and our eyes, just as if it were a photograph that we carried from the television screen to our eyes. But this is not how it works. Instead, nature manipulates the image in the frequency domain. A filter is applied to the image that modifies the properties (called the *phase*) of every one of those 2,073,600 plane waves making up the picture. If this is done in just the right way, the image will appear to have moved from the screen to our eyes. The main point is that this image processing doesn't happen in the space between the screen and our eyes. It happens in the frequency domain. Where is that? It's everywhere! It's amazing that you must use the timeless and spaceless physics of patterns and relationships to understand even this basic phenomenon that happens every time you open your eyes. Wholeness appears right in front of us at every moment.

Throughout this book, we have used the concept of filters to describe how we adjust our feelings and reactions to experiences in daily life. Here we see where that concept comes from. In the city image, we see how each frequency "component" adds an essential quality to the whole image. Earlier I made the analogy that our own personal qualities work like this too. We as a whole are made of many components. Each is essential and applies everywhere. If we carry unhealed resentment from a past relationship, that resentment shows up everywhere in our lives, just like any single frequency component in a photo affects the whole image. This is an analogy, not an equality, yet it provides a useful way to think of our personality and our conditioning as filters on our natural wholeness.

Wholeness in Digital Photos

Now we can look at some tangible and intuitive examples of holism in digital media. Holism originates from the all-to-one and one-to-all relationship between regular space and the frequency domain. Adobe Photoshop uses this, as does Snapchat and any software that performs filters on visual imagery. This applies to audio as well. Anytime your voice is converted into text or the music on your stereo is modified with a graphic equalizer, changes are made in the frequency domain that affect *all* of the original sound file.

Picture the wholeness of a s'more. That's right, the kind you roast around a campfire. A s'more is a dessert made of chocolate, marshmallows, and graham crackers. There is flavor information—the "s'more-ness"—diffused throughout it. If you separate them, you no longer call it a s'more. The special taste of a s'more comes from the relationships between the ingredients. If you taste the separate parts alone, its essence is lost.

We can see the same wholeness in a digital photograph. Starting with the photo of a city in Figure 13.11, the photo can be converted into frequencies using the Fourier transform, the mathematical process we discussed earlier.

The result is its "spectrum" in the frequency domain, shown in Figure 13.12. This image is a bit hard to grasp, so let's walk through it. Each dot tells the computer to plot a pattern like in Figures 13.1 through 13.6. The form of the pattern depends on the dot's angle and distance from the middle of the page. The dot's strength tells the computer how dark to make the line.

FIGURE 13.12. We saw in Figures 13.7 through 13.11 that the city is actually made of thousands of simple line patterns, each represented by a dot in the frequency domain. (The dots are white and the background is black.)

Together all these patterns make up the original city image. Although the buildings don't exactly look like they can be reduced to a bunch of simple patterns, nevertheless there is hidden regularity. Figure 13.12 is a complete picture of the city but drawn in the language of the frequency domain!* The Fourier transform picks out every pattern present in the picture, no matter how subtle.

We can draw the city from its spectrum using the frequency domain, as we did in the last section, but this time emphasizing the difference between low frequencies and high frequencies. Figure 13.13 shows the spectrum of the image with all the regions outside the center blackened with a "mask." The mask removes some frequencies, like black tape on a window.

In Figure 13.14 we see the magic of the frequency domain. Only vague forms can be seen in the image. These are the low-frequency components of the image. They paint the canvas with broad strokes.

---

* For readers familiar with the Fourier transform, you might point out that this particular image is the magnitude of the spectrum, so the phase information has been removed. This is a detail in graphics, not in essential points!

Figures 13.15 through 13.22 gradually increase the amount of low frequencies that we include. In each image we see more detail appearing in the photo.

FIGURE 13.13. Here is the spectrum from Figure 13.12 but only the lowest frequency values. The black represents the value zero, so this is called a *mask*. The high-frequency values are omitted.

FIGURE 13.14. With only the lowest frequencies, we see just a vague impression of the image.

FIGURE 13.15. We gradually include higher frequencies in our drawing.

FIGURE 13.16. With higher frequencies included, we start to see some of the form of the city emerge.

FIGURE 13.17. We are steadily increasing the number of included higher frequencies.

FIGURE 13.18. Now we really see the shape of the city, but there are clearly some artifacts, echoing the shape of the city buildings. They are introduced into the picture because of the missing high frequencies.

FIGURE 13.19. We are including even more high frequencies but still not close to all of them.

FIGURE 13.20. Even without including all the high frequencies, which correspond to the details of the scene, we see an image that looks excellent.

FIGURE 13.21. Here we test what happens when a small portion of low-frequency information in the frequency domain is blackened.

FIGURE 13.22. Blocking out just a small region of the frequency domain has the effect of creating artifacts throughout the entire image. Each portion of the frequency domain affects all of regular space. If we could alter the frequency domain here on Earth, could we affect planets and stars far away from us?

Figure 13.13 and the others with black frames are missing some information. Part of the frequency domain is blacked-out. If we drew the city pixel by pixel, having a blacked-out frame would mean we could only draw a portion of the city. If I sent you half of a photograph, that's all you'd get.

But instead of a brush that paints pixel by pixel, Fourier's holistic brush paints pattern by pattern. Each "brush stroke" covers the whole canvas. When we haven't applied all the brush strokes, the drawing may be blurry or have wavy artifacts, shapes, and patterns that shouldn't be there (see especially Figure 3.18). These artifacts are not part of the original drawing, and they diminish as we add more and more frequencies.*

Let's hammer the idea home with one more example. We can add a filter (or mask) to the spectrum as in Figure 13.21. An eraser was used to remove a small amount of data in the middle of the spectrum. This, it turns out, means that I have removed some (a very small amount) of the low-frequency information from the image. You can think of this as removing all the patterns that repeat slowly, while keeping the patterns that repeat quickly.

Converting this back into regular space, now using the *inverse* Fourier transform, we see in Figure 13.22 that the entire image of the city has been affected. The artifacts are plain. We can easily see the slowly varying shading that corresponds to the low-frequency data that we removed. *By removing data from only a small region of the frequency domain, we have painted the entire original picture with new brush strokes.* We can see here how each little part of the frequency domain corresponds to the whole city.

Wholeness in Digital Music

This behavior exists for music as well. Music is a pattern of waves in time, rather than space, but the story is the same. In photos high frequencies capture the fine details, but in music high frequencies capture the high-pitched notes. If we convert a song into the frequency domain, we can erase some

---

* For hologram lovers, note that this is why you can cut a hologram in half and still see the entire hologram. All you need to do is change your perspective on the hologram and you see the image that was blocked out.

of the high-frequency sounds because human ears can't hear it anyway. But if the procedure is not done carefully, there will be artifacts in the audio quality, extra patterns of sound that warp the music when you play it back.

The MP3 digital music format was developed to optimize this process. Through careful application of the right filters in the frequency domain, an MP3 is made ten times smaller than a raw sound file recorded from a microphone.

It's fascinating to note that when we convert the file to the frequency domain it no longer has a beginning and an end in time! It is timeless, containing only a map of the frequencies. We could not play it on a stereo in the frequency domain format. But the information about time ordering is only hidden, not lost. It is encoded into the relationships between frequencies, so when we perform the process in reverse the music gets put back into proper time ordering and we can listen to it again. To me, this is the most amazing example of magic in science.

This leads to some very useful techniques and tools, such as a graphic equalizer. Let's say we are recording a symphony, only to realize afterward that our camera had an internal buzz at a frequency of 1000 Hz. Is our recording ruined? No! If we graph the data in the frequency domain, we will see a bump around 1000 Hz representing that unwanted buzz. By erasing just that little bit of data and converting the file back to proper time ordering, we can play back the symphony without buzz. Just as we did with the filter and image in Figures 13.21 and 13.22, we have altered the entire song with just one small change in the frequency domain!

However, the design of the MP3, while awesome, is actually a hack. We don't do it quite as I have described. You see, humans are very attached to time. We don't know how to make sense of a data file that doesn't have a beginning, a middle, and an end. We want to think in terms of a linear progression of time, a string of time-ordered moments.

So we hack it. We chop up the original sound file into short sound bytes. Then we Fourier-transform each sound byte to get the frequency data at that moment in time, keeping the sound bytes in the original time order. In other words, we force the file to stay organized by the clock, even in the frequency domain.

Regardless of this practical consideration, the important point remains: both images and sound can be converted to the frequency domain, into a form that doesn't have any space or time ordering. This provides us a very tangible example of the wholeness captured by the Fourier transform and the one-to-all, all-to-one relationship between regular space and *pattern space*.

Meditation: You Can't Touch without Being Touched

Imagine yourself standing in a meadow. Above you is the evening sky with the colors of the sunset. In front of you are some reeds at the edge of a lake. You walk forward to a gap in the reeds and step out onto the water. It supports you as you walk slowly past the reeds toward the middle of the lake. You watch your feet as you walk; each time your foot touches the water, ripples spread outward from the point of contact. The ripples form concentric circles around each foot, forming a web of circles from all your steps.

You get to the middle of the lake, and you stop walking. You glance up at the sunset and the darkening sky, with a few first stars shining in the deep blueness. You take a deep breath. You are curious about the ripples on the water. You direct your eyes downward at the water again, to find that all the ripples have died away. The surface of the lake is like a mirror; you can see three stars reflected in it. You bend down, and ever so carefully extend your index finger toward the surface.

The instant your finger makes contact with the water surface, ripples emanate outward from the spot your finger touched. You pull your finger away, and another set of identical ripples is created. You gaze softly in wonder at this ever-so-familiar pattern. It is like the pattern of raindrops on the sidewalk, or craters on the Moon.

You extend your finger even more carefully to the water surface. When you touch it, yet again ripples race outward from your finger across the lake. Just by the slightest interaction with the lake, you have caused ripples. You marvel that the ripples are the water's response to you. There is no half-hearted interaction with the water surface. It is all or nothing, and the ripples are the undeniable evidence that you were here on the lake.

You raise your gaze and turn around to see the Moon rising above the edge of the lake. Slowly and gracefully you walk across the water back to the shore. You step onto the soft grass and the firm earth. You feel yourself interacting with the wondrous Earth in a thousand simultaneous ways and give thanks.

Wholeness in Standing Waves

I've long been fascinated with standing waves. Unlike waves traveling past you, such as on the ocean or on the surface of a lake, standing waves stay in place. When traveling waves bounce back and forth in just the right way, they interact and form a fixed shape in space.

For instance, the waves on a guitar string are standing waves. They fill up the whole string and vibrate with a certain frequency to give a unified tone. We might call the string "an E string" because a standing wave of 330 cycles every second lives on it. This description doesn't apply only to the left, or right, or middle of the string. The E tuning is a property of the entire string as-a-whole. The standing wave is not at a specific point in the string. It is a phenomenon that exists everywhere along the string, tying it together and unable to be divided.

A laser is another even more amazing example of standing waves. Imagine trapping light in a tube between two mirrors. It just keeps bouncing back and forth. The light between the mirrors is like a wave, sloshing this way and that. If we adjust the length just right, we can make sure that each time the light repeats its round trip it reinforces itself. The high points in the wave get higher, and the low points get lower.

Suddenly, instead of a tube of randomly oriented light, our tube is filled with highly organized light, laser-light, shining in sync with itself. It's like when, as a kid, you moved your body backward and forward in the bathtub, so that the waves of bathwater turned into a tidal wave.

This process of light amplification is called *lasing*. The light forms a standing wave inside the tube with mirrors, so that all the light in the tube takes on a collective behavior. To get a laser working, one has to create the conditions for lasing. The transition from random, uncoordinated light takes place all-at-once. The standing waves that are set up between the endpoints of the laser tube are properties of the whole tube, and they "turn on" all of a sudden when the conditions for lasing are met.

Wholeness in the Pinhole Camera

If you took a physics class in high school, you may remember a simple experiment called the *pinhole camera*, also known as the *camera obscura*. Ibn al-Haytham is credited with its invention around 1000 AD. You can try it at home. Place cardboard or paper with a very small hole in it (the aperture) near a brightly lit object. If there is a wall or a white piece of paper to serve as a screen on the other side, you can see an upside-down image of the original scene. If you place a piece of film there, you could capture a photograph of the scene. This camera existed long before lenses were invented.

You can see this effect in an eclipse of the Sun, too. The leaves of trees form tiny pinholes of sunlight, and on the ground you will see little images of the crescent Sun. In fact, when you look at the shadows of trees, you are always seeing little images of the Sun. Because the sun is round, those little images of the Sun on the ground just look like overlapping round bright spots, so it seems like you are just looking at a "simple shadow" of the leaves. But during an eclipse, when the shape of the sun is not round but crescent, you realize you are looking at thousands of little images of the Sun. The effect is stunning.

Meditation: The Pinhole Camera

Imagine you are standing in a clearing in the woods. On all sides, maybe fifty feet away, are beautiful trees of various shapes and sizes. They are blowing gently in the wind. You feel the breeze in your hair. It fills you with a feeling of spaciousness.

You look down in front of you to see a cardboard box on the grass, about the size of a microwave oven. A pair of scissors is on the ground next to the box. You pick up the scissors and cut a hole just big enough to fit your head through. You see a pushpin lying there. Turning the box right side up, you push the pin into the other side of the box and remove it, leaving a tiny hole for light to get in. Slipping the box over your head, you are enveloped in complete darkness except for the tiny pinhole. Instead of looking directly at the pinhole inside the box, you turn the box to point your face away from the pinhole.

It's totally dark, but as your eyes adjust, you notice light is shining through the pinhole behind you and onto the inside of the box in front of you, like a movie screen. What you see is not just a splotch of light ... you can make out shapes. In fact, you see what appear to be trees. The images are very dim. The pinhole is only letting in a small amount of light. As you look more carefully, you notice that the trees on the inside of the box are upside down. You can see them rustling in the breeze, with their trunks pointing up and their branches pointing down. You are seeing an upside-down image of the trees behind you on the dark inside of the box in front of you.

With the pin, you poke another hole in the back of the box, very close to the first. On the screen you see another complete image of the trees. The new scene is slightly out of place from the first scene, creating a "double-vision" effect. The two holes create two side-by-side images.

Now you use the pin to merge the two small holes into one big hole. Instead of double-vision you see one blurry image of the trees. You realize that the wider hole is like many tiny pinholes smashed together. The blurry image of the trees is like many crisp images of the trees smashed together.

Now you remember you have a special healing salve in your pocket. Taking a little on your finger, with your head still inside the box, you gently rub the salve into the outside of the pinhole. The box starts to heal and the pinhole shrinks. You notice the blurriness going away and the image of the trees getting sharper, but the image also grows dimmer as the pinhole gets smaller.

Your eyes are so well adjusted that you can still see the image as the pinhole gets really small. It gets more and more detailed in its resolution. You can make out sharp details on each tiny leaf that you couldn't see before.

The pinhole gets so small that only one photon of light appears on the screen at a time, yet wherever the photon lands, it adds the right color to the image. Amazingly, each photon passing through the pinhole contains information about the entire map of the scene outside the box. With every photon comes a blueprint for every detail of every tree, cloud, and blade of grass.

You immediately understand that the world you normally see is not a bunch of separate pixels like an LCD TV screen. Each photon contains the entirety of the world within it, and then becomes a bright or dark pixel based upon its relationship to the whole.

You gently pull the box off of your head. You blink at the brightness all around and feel immersed in the holographic bath of light that has always surrounded you. Every inch of your body is being bathed by photons containing details of the entire scene around. You breathe in deeply. You absorb the wholeness of your surroundings within which you are a perfect fit.

When I was a physics teacher, I did this pinhole camera experiment inside a cardboard box during a professional development program created by the Exploratorium science museum in San Francisco. The profundity of this experiment hit home like it never had before, and it continues to inspire awe. The smaller we make the hole, the better we see the image. The entire image of the surrounding scene, what I had assumed was a billion independent photons arranged like the pixels of a TV, was captured in this tiny pinhole. Could each photon that gets through the pinhole contain a whole, detailed image of the surroundings?

I had been taught that a photon, or particle of light, was very simple. It was defined by its speed, its wavelength—visible light being about 0.01 millimeters, and microwaves about 1 millimeter—and its frequency. But here I was being shown that the complex relationships of color and brightness across an entire, macroscopic scene (like the trees in the sidebar) seemed to go all the way down as small as one could go. No matter where you placed the pinhole, it would capture the entire scene. The image as-a-whole was contained in the single part.

The single photon getting through the tiny pinhole can become any part of the final scene because *all of the frequency domain exists at every point in space*. In other words, even in the most minute gap in the box exists the entire spectrum of the outdoor scene. From this complete spectrum can be reconstructed the entire image in space when it strikes the other side of the box, much like re-creating an entire organism from a single strand of DNA.

The pinhole camera is one of the most beautiful, accessible experiments we can do with light. It demonstrates clearly that the whole is contained in its entirety within each part, pulling the rug completely out from under the reductionist argument. And you can create it in your own living room with a third grader.[3]

Wholeness in Lenses

We can see the inherent wholeness of light in a device very familiar to all of us: the lens. Many centuries after the invention of the pinhole camera, the theory of optics was developed and the lens was invented. A lens has many of the same properties as a pinhole, but it can be manipulated and customized. Lenses exist in our eyes to bend the incoming light into focus at the back of our eye, where it lands on the retina. The retina is like a screen or piece of film that captures the focused light. Lenses are also used in cameras to form an image on film. The lenses in a camera can be warped, stacked, and adjusted to manipulate what you see. Lenses can be used to focus a laser beam to a tiny point that can read the reflective pits on a DVD.

Let's say we set up our lens to capture the light from trees in front of us. The lens captures all the light from its surface and focuses it to one point: the *focal point*. If we place a screen at the focal point, it will show us a sharply focused upside-down image of the trees. When you look at the trees with your own eyes, a map of the trees appears on your retina, a miniature version of the real world in front of you. This is called a *one-to-one map*, because each point on the real trees corresponds to a point on your retina.

But the situation is not as simple as it first sounds. When light leaves the trees, it spreads out in all directions, similar to our discussion on the propagation of an image from your television screen. The light that reaches the lens has been transformed into *pattern space*, and the lens's job is to undo this. The lens performs a Fourier transform. Every part of the light striking the surface of the lens gets mixed together and lands at every part of the focused image. The top of the tallest tree, for instance, is just one point in the final image that is made of all the light hitting the lens. The same is true of the bottom of the trunk. In textbooks we are shown a diagram of a lens focusing all its rays to a single point. But this is a misleading simplification. The lens is doing that for every point on the image, as shown in Figure 13.23. The standard lens diagram found in textbooks is really a hint of the all-to-one and one-to-all relationship of the frequency domain and regular space, only the diagram is simplified in a way that explicitly overlooks this amazing feature. It is, for the very reasons it is interesting to us in this book, very hard to draw!

You can test this yourself. If you wear glasses, you may have noticed that when small amounts of dust fall on your glasses you can still see through them. Or if you walk up very close to a dusty window, you hardly notice the dust that is mere millimeters from your eye. Why is this? An optician would say it is because the light hitting the dust and entering your eye is being focused to a point very far away. The muscles in your eye are not able to deform the lens enough to bring such close objects into focus on your retina. What your retina sees is therefore an unfocused image of the dust.

This is correct, of course, but if we are satisfied with this, we might miss something interesting. What does the optician mean by "unfocused image"? An unfocused image is simply the "all" portion of the one-to-all and all-to-one relationship between the frequency domain and regular space. Since the dust is so close to your eye lens, it spreads out on the way to your retina but doesn't get refocused by the lens. This spreading out is the light's holistic nature. If the dust were farther away from your eye, the light bouncing off of it would spread more *before* it hit your lens, and the lens could refocus it onto your retina. You would then see the dust clearly as a single image. Instead, that light spreads across your entire retina, and no clear image is formed.

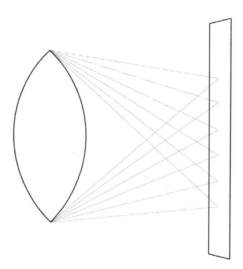

FIGURE 13.23. A lens captures an all-to-one and one-to-all relationship. It Fourier-transforms the light by sending light striking every portion of its surface toward each point on the screen. The light gets mixed up in such a way that a meaningful image becomes unlocked from inside the light wave.

So when you have dust on your glasses that you can't see, you *are* in fact seeing the dust. It is just spread over your whole field of view! It is mixed in with all the other light from every other region, and you don't know that you are perceiving it at all. Your overall view is dimmer, because a little bit of the overall illumination has been removed. But you hardly notice that.

You can try this for yourself, being very careful not to hurt yourself. If you bring something very thin up to your eye (not too close!), you will find that you can look right through it. The temples of my glasses (the long part that goes over my ears) work safely for this. If I take off my glasses and look at that thin part up close, I can see right through it! There is a scene in *The Empire Strikes Back* in which Han Solo and Luke Skywalker are evading Darth Vader. They are in the Millennium Falcon, trying to escape a massive *Star Destroyer* spaceship. To avoid the radar, Han Solo docks their ship on the underbelly of the enemy ship, so close that the radar cannot detect them. The radar light bouncing off the Millennium Falcon is like the dust on your glasses. It is too close to be resolved by the radar detection device.

The light striking our eyes has integrity as-a-whole. It is being spread everywhere and cannot be broken into pieces. *Every* point we see in an image depends on *all* of the light striking the lenses of our eyes. We experience the mysterious wholeness of light everyday, we are immersed in it, and we are mostly unaware. We go about our day completely convinced that objects are fundamentally separate from each other, that there is a one-to-one, reductionist map between points in space. Maps help us function in the world. They help us locate our coffee on the desk, or the location of our computer in front of us, or the ingredients we are using to make breakfast in the kitchen. A map forms in our mind, relating our own body to the world around us. This works great because the map accurately shows us where those objects are when we reach out to grab them. Yet there is a hidden wholeness that is obscured. The mapping of the world that we make in our brain is so effective that we think space itself really is this way. We jump to the conclusion that the fundamental nature of the universe *is* as it *seems* to us, arranged in a grid whose coordinates can be labeled and dissected. But for light, this is not so at all. The space that light travels through cannot be dissected. Space exists as-a-whole. All of the light striking a lens is part of one integral whole. Entire regions of space have an *identity*. The light within that region cannot be separated. One edge cannot be isolated from the other.

As we shall see in chapter 14, this wholeness has no bounds. Light leaving a star travels outward into space. But it does not travel like a bullet in a straight line. It expands as a wave of possibility in many directions. In some models, it is thought of as a standing wave in space *and time* so that its start and end points form a single, holistic region. So for light leaving the Sun, which takes eight minutes to travel ninety-six million miles to Earth, this region of space and time around the Sun exists as-a-whole. We could say poetically that this region of wholeness is ninety-six million miles long and eight minutes wide.

But in physics, this region is called a *null interval*. Why? When one considers that light travels at about 11 million miles per minute, one finds that the *speed* multiplied by the *time* taken exactly cancels the distance traveled. For light, the true measure of separation between Earth and Sun—what's called the *Lorentz interval*—is zero. It is as if the beginning and the end points of light's travel are the same.

This wholeness extends not just eight light-minutes into space but as far as the eye can see. We are part of a realm of wholeness that extends to the Andromeda galaxy. The realm of wholeness extends even as far as the observable universe, for it is light itself that defines the extent of the observable universe. All forms of light behave this way—radio waves, X-rays, visible light—so the result is quite general.

It is easy to believe that the true nature of the space we live in is divided into separate sections. Here on Earth we are separate from the Sun and all the other planets in the solar system. Here in the United States I am separate from you in China or Italy. But this is only part of the picture. The space we inhabit is not divided up like this. It is part of one whole extent, with all its parts connected through the mathematics of the Fourier transform and the underlying frequency domain, illuminated for us through the well-documented and easily witnessed behavior of light. We've looked at the holographic or whole nature of ripples in a pond, a pinhole, a lens, and light raining down from space.

Phenomena that you take for granted every day serve as a gateway to this unfamiliar way of looking at the world.

Let's continue this journey into everyday wholeness with more examples from science, technology, and the real world.

Wholeness in the Brain

Some of the examples so far are familiar to you, others maybe not. But *pattern space* is not foreign to you, even if this is the first time you have read about it. You naturally recognize patterns. Leading theories of the brain hold that you use the Fourier transform to convert the words you hear or the visual scene in front of you into pattern space. You can readily see this for yourself using the simple technique in the sidebar. In pattern space, a small adjustment of your finger's position in regular space causes all of pattern space to change.

How does this work? Imagine a very complicated spreadsheet tracking your budget. If you change one number at the top, every number in the spreadsheet may change. Similarly the displacement of one small portion of your body affects the properties of every part of pattern space.

Because of the all-to-one and one-to-all relationship, no part of pattern space is unchanged by even the simplest shift of your finger. Remember, pattern space is about patterns and relationships. When we move our finger relative to our hand, the relationship changes throughout.

If your brain scanned your field of view pixel by pixel to find evidence of motion, it would be mind-numbingly slow. But if your brain is scanning in pattern space instead, it can notice the *difference* easily, for the whole spectrum changes with just the slightest adjustment to the scene.

Karl Pribram was one of the first neuroscientists to eloquently state the hypothesis that the brain operates holographically (or "holonomically").[5,6,7] Pribram wrote, "People who have strokes or head injuries of any kind never lose a particular memory trace. Memory is all of a piece; it seems to be distributed in the brain so that even huge destructions do not remove a particular piece of a memory."[8] This enhanced resilience of the brain provides an obvious evolutionary advantage.

Seeing Motion by Wiggling Your Finger

The brain appears to operate as-a-whole. To see this, sit comfortably in your chair and stare gently straight ahead with your hands in your lap. Keep your eyes fixed on something directly in front of you. Now notice what you see in the periphery. As you focus straight ahead, objects in the periphery should be visible but not sharply focused. Now place your hands in your lap just within the periphery. You should be able to tell they are there but unable to make out many details.

Of a sudden, move one finger just an inch or so. While the finger itself was not in focus, its movement is quite obvious. If you couldn't tell before which one was your pointer finger, you can easily tell after you see it wiggle in your peripheral vision.

It is not the finger but the movement that is picked up by your brain.

Karl Pribram provides another example of the brain's focus on differences. "(If) you wiggle a little bit ... you realize that you have got clothes on, but most of the time you do not notice them at all."[4]

The tendency of your brain to detect pattern also has an obvious evolutionary advantage in nature. Imagine yourself a hunter in the wilderness or a prehistoric sea animal with a newly evolved eye. You stare at your surroundings—the grass and trees, or the open ocean—and everything seems serene. But suddenly motion catches your eye and you jerk around to see the prey or predator near you. You couldn't see them until they moved, but your quick response gives you the jump, allowing you the optimal chances of survival.

Neuroscientist Fergus Campbell found evidence that the eye responds not to individual objects but to spatial patterns within its field of view.[9] "Fergus Campbell's work ... provided the first evidence that the brain might indeed function very much like a hologram ... If the brain is made not to construct stick figures, but to resonate with particular frequencies there is a much richer perception that is like the richness of sound a pianist can produce from a piano," Pribram writes.[10] He is emphasizing that the brain doesn't draw things from top to bottom in space, like we would draw a stick figure. Rather it constructs its images by layering on high- and low-frequency patterns, like is done in photo-editing software and shown in Figures 13.7 through 13.11.*

The anatomy of the brain thus suggested was quite different from existing models. "Essentially, the theory reads that the brain at one stage of processing performs its analyses in the frequency domain. This is accomplished at the junctions between neurons not within neurons," Pribram says.[11]

These ideas are central to Ray Kurzweil's modern model of brain function. Kurzweil's background in audio signal processing is a natural prerequisite for expanding upon Pribram's holographic model of the mind.† In his book *How to Create a Mind*, Kurzweil's model of the brain is motivated by the way humans think in patterns rather than individual bits of information. For instance, we can recite the alphabet very easily,

---

* You can convince yourself that something fishy is going on for your brain simply from the dizzy feeling you get from looking at Figure 13.1. Shift your head around a little bit. Your brain adds artifacts—other patterns—into the straight lines. (Imagine for a moment that you have someone translate this book into German for you, but you surprise them with one page of the book *already in* German—they would be confused for a moment, no?) If I'm not mistaken, your brain is trying to do a Fourier-transform on this image, but it's already in the Fourier form. Your brain is confused by it!

† I first became a fan of Kurzweil when I was a teenager. One of Kurzweil's first companies made electronic pianos. They were the most realistic-sounding products on the market, based upon exquisite sound modeling in the frequency domain, of the sort we are talking about here. I desperately wanted one of these that I could bring with me to gigs instead of the clunky plastic synthesizer I had. A synthesizer constructs sounds from scratch, generating a limited set of frequencies to construct a false rendition of a piano. Kurzweil's keyboards were based on "sampling," or capturing the actual spectrum from a real piano and reproducing it in real time as the performer struck the keys. The sound of the Kurzweil keyboards was head and shoulders above any of its competitors.

but if we try to do it in reverse order most of us are quickly stumped. We memorize the pattern of letters as-a-whole, not the individual letters in isolation.

Similarly our thoughts pass through our mind in a relentless train from one to the next. You may be having a conversation with a friend and get off on a tangent, and suddenly ask yourself out loud, "Now, why did I bring that up?" Your friend says, "We were talking about my trip to the store yesterday," and you reply, "Ah yes, now I remember … " and finish the story.

Kurzweil points out that our behaviors are also pattern-based. It is quite easy for us to brush our teeth even while paying attention to something else. Getting the toothbrush wet leads naturally to unscrewing the cap on the toothpaste, which is inevitably followed by squeezing the toothpaste onto the brush. This sequence of events is a pattern of patterns ingrained into our nervous system, and I often do all this easily while simultaneously reading a book!‡

We are also extremely good at recognizing faces. Once the pattern of someone's features is imprinted in our mind, we can recognize a face from only a partial glimpse, in the same way we can recognize a well-known song from only a few notes. With both faces and songs, the partial pattern triggers the recall of an entire pattern stored in memory. This is the same reason that we can sometimes remember someone's name more easily by recalling some quirk about them.

We can also consider the evidence from both personal experience and laboratory experiment that memories fade over time.[12, 13, 14] This is consistent with the idea that they may be stored as patterns in a network of neurons rather than as bits in a single neuron. If memories were stored as bits of information, either on or off, it would be hard to understand how a memory could fade. But if a pattern becomes polluted with extraneous noise, it becomes harder to detect.

Kurzweil models the neocortex, which is responsible for abstract thought, in terms of "pattern recognizers." From knowledge of anatomy

---

‡ The hardest part is not the brushing of teeth but keeping the book open with one hand!

he estimates that there are about 300 million pattern recognizers in the brain.[15] To master a skill, our brain tries to recognize patterns, and it is estimated that an adult can reference about a hundred thousand patterns simultaneously to solve a problem. Kurzweil's pattern recognizers are more than plentiful enough to handle this. Note that if each pattern recognizer stored one *bit* instead of one *pattern*, it could only store one photo in TIFF format!

Memory, like light, is not reducible to the sum of its parts. Rather, there is a wholeness to the thoughts we have. Both in the brain and in the physics of light, wholeness arises because the information exists in *pattern space*.

Pribram and others recognized the importance of these findings beyond concerns about the wiring of the brain. He wrote,

> The importance of holonomic reality is that it constitutes what David Bohm calls an "enfolded" or "implicate order" … Everything is enfolded into everything else and distributed all over the system. What we do with our sense organs and telescopes … is to unfold that enfolded order … We owe to David Bohm the conceptualization that there is an order in the universe … which is spaceless and timeless … We now find that an important aspect of brain function is also accomplished in the holonomic domain.
>
> (In this domain) the ordinary Euclidean and Newtonian dimensions of space and time become enfolded. Synchronicities and correlations characterize the operations occurring in this domain. There is no here, no there. There is no-thing. But this holonomic order is not empty; it is a boundariless plenum filling and flowing. Discovery of these characteristics of the holonomic order in physics and in the brain sciences has intrigued mystics and scholars steeped in the esoteric traditions of East and West: for is not this just what they have been experiencing all along?[16]

In the next chapters, we will look at how the Holographic Paradigm allows us to see more connections between the events in our lives. Synchronicity and flow fit well within the Holographic Paradigm. Time and space themselves are whole, and our activities within them have ripples that exist spacelessly and timelessly through the world and through history.

14

THE MECHANICS OF THE HOLOGRAPHIC MULTIVERSE*

Making good choices is difficult. Whether they are in our business, our social lives, our finances, or our family, the choices we make determine the direction of our lives and our society. In the West we believe in a narrow path of cause and effect. To understand why something has happened, we look for the physical actions that precipitated it. We do not often look at the entire scene; we do not look for the context within which our experiences happen.

When a job loss happens, or when illness strikes, or when a new client appears on our doorstep, the question arises as to whether these events are meaningful. Did they happen for a reason? This is the same as seeking to understand the context in which they have occurred. What are all the environmental elements that help me see where this experience falls in the broader span of my life?

---

* Some ideas in this chapter are original or not yet well accepted in mainstream physics. They are in flux and may change as they undergo peer review. I encourage you to read with both curiosity and skepticism. For background on the history of theories about the holographic multiverse, see Appendix C.

My quest to understand synchronicity arose from the spooky timing of certain events in my life. A few years ago, I had an experience of this sort of timing that struck me as meaningful. I had been working for a technology company for a number of years but contemplating what it would take to start my own business. I wanted to actively pursue research, writing, and lecturing on synchronicity and its connection to physics. Driving home in traffic one day I called my coach and wrestled with the idea of leaving my job. There were pros and cons, and although I had been doing the research and lecturing work for a while on the side, it was a scary idea to leave the safety of employment and accept the unknown of committing to that path. I said to my coach, "The truth is, I don't have any speaking engagements or other specific opportunities pulling me away from my job or causing a conflict. I have plenty of flexibility to work on my business in parallel to working at the technology company. Until there is a conflict, I guess I won't worry about having to choose." I hung up the phone and felt relieved that I had some clarity, and that there were no hard decisions to be made just yet.

A few days later, some very unusual events started to unfold at work. A friend of mine, a well-liked senior manager, announced he was leaving the company, and this precipitated a shake-up in the leadership. Our team was now rudderless. Out of the blue I was approached by the president. She said, "I know you are happy with your current arrangement, but it would be really great if you were interested in a leadership role on the team." I was in shock and somewhat excited by her invitation, even though it would mean a busier schedule and more responsibility. I responded that I would be open to considering it.

I was promoted to a coordination role for the engineering team. I found myself much more engaged at work, enjoying the challenges and working much longer hours. I was the main contact point between the sales team, the customer service team, and the engineering team as we navigated an onslaught of new projects. I no longer had spare time to devote to research or writing on the side. It occurred to me that my lack of a decision was, in fact, a decision. Spontaneous events had unfolded, which made the decision for me.

I enjoyed the new job, for a while. A year or so later I was handling one of our biggest clients and had an unpleasant interaction with our main contact at the company. I realized that I was unhappy with the direction the project was going, and that it was clearly going to get worse before it got better. Now my decision was clearer. In order to live the life I really wanted, I would have to leave the company.

My manager was disappointed with but supportive of my decision, and the transition went well. Soon I found myself applying to graduate school, and a new dream (or the reviving of an old one) was in motion. The flow of events here was not one that I could have predicted. It allowed me to experiment with the options I had in front of me, even over the course of an entire year, and eventually come to a choice that felt right. I couldn't tell in advance what would feel right. In retrospect I see that my fear of striking out on my own made me unable to choose to leave the company the first time, but after these events had unfolded, I felt ready for that.

We can handle difficult choices better by becoming more aware of the connection between our inner world of thoughts, feelings, and decisions and our outer world of actions, opportunities, and obstacles. This reflects the Theory U put forward by Otto Scharmer in his book *The Essentials of Theory U*. The tops of the U represent the circumstances we currently have and the circumstances we want. The bottom of the U represents the inner landscape that we must traverse in order to successfully navigate to the new circumstances. Rather than compartmentalizing, we see our life as one whole stage of interconnected events. There was no causal link between my decision to not leave my job and my friend's decision to quit. Yet the timing of these events was, in retrospect, perfectly aligned to help me through a difficult process of growth.

My coach, myself, my friend, and the company president are all part of a whole scene that provides context for my experiences. Through the process of meaningful history selection described in *Living in Flow*, I am correlated to these other elements in a whole network of contextual influences. Instead of thinking of our environment as a bunch of disconnected, irrelevant props, by taking a perspective of wholeness we can see the wider patterns emerging. This perspective helps us see how one event in our life provides context for other events, how getting fired from a job may be the right launching pad for a better experience, or how an illness may serve as a useful experience for deepening our relationships.

The art of choice is to see the wholeness in life experiences and understand which growth opportunities we are inviting with every decision. By experiencing the different aspects of our lives as part of a whole—family, work, personal endeavors—we are able to see paths available to us that may have been otherwise hidden.

Viewing our environment in its wholeness shouldn't be considered a fringe concept, even if it stretches our notions of physical cause and effect. People from every walk of life have experiences of this nature, feeling awe over the surprising order with which life events unfold. Astrophysicist Neil deGrasse Tyson relates a story about visiting the Hayden Planetarium as a kid, where now, many years later, he is the executive director. The experience had a pivotal effect on his career path.

> My first visit to the Hayden Planetarium … it was a matter of exposure for my brother and sister and me. You don't want your options to be limited, when you're asked what do you want to be when you grow up. And the more things you see as a child, the more options you have to reach for, if something piques your interest. And for me, my first visit to the planetarium, I'm convinced, in fact, that it was the universe that chose me.[1]

Tyson presents himself as a very reasoned and fairly conservative thinker in the world of physics and in the public sphere. He takes a no-nonsense approach to science, politics, and society that I appreciate. Is it unreasonable for him to think that the universe "chose" him? No, I don't think so. I imagine he is simply referring to all the many timely opportunities that came to him as his academic career progressed. But I suggest that those experiences cannot be explained solely within the standard framework of causality. One must invoke some broader organizing principle, and although he might or might not agree with the way I have stated it, his choice of words seems to indicate agreement that something more is needed. We need an understanding of the flow of events in our lives that leaves room for the interconnectedness of our choices.

Space, Time, and Synchronicity

It may be surprising that a field as abstract as physics may help us better understand something as intimate as our personal choices. Meaningful history selection provides a possible explanation for synchronicities within the framework of physical science. A synchronicity is a chance event that feels meaningful because it leads to an experience you are seeking to have. Remember the woman who was asked by her manicurist for a stamp card, so she reached in her purse to find the business card with the information about her blue topaz ring? Synchronicity is the way nature weaves our lives together. It is the way the cosmos responds to our choices. Any event can be described through the time and place where it occurs, so the study of meaningful events is intimately related to our understanding of space and time.

What is space? What is time? You may be surprised to know that the very questions that Galileo and Newton pondered are still pondered today. One of my favorite aspects of physics is that even 400 years after the discipline was established, we are still working on these basic problems. As physics students, one of the first things we'll do is investigate space and time by doing experiments that measure the motion of rolling balls. How does the position of the ball change in time? Yet by the end of our studies, after we have covered thousands of hours of intervening material, we end up back at square one. Loop quantum gravity and string theory, for instance, are both theories attempting to understand space and time at a more fundamental level. Thanks to Newton, Einstein, and scores of other physicists, we have progressed very far in our understanding of these fundamentals. But in some ways, we still don't understand the basics. In some fields, say chemistry, progress is denoted by more and more complex calculations and theory. In physics, the more advanced we get, the more basic the subject of inquiry gets.

So the study of how events get woven together in our lives is both unresolved and profoundly important. Even if you do not understand or have interest in physics, it is useful to understand what we do and don't know about space and time. We operate in space every day. Our life is made of the flow of time. If we go about making choices with unfounded assumptions or worldviews, we are bound to miss the subtleties and wonder, and possibly even hamper our progress through life.

Have you ever been told by your Elders that time flies? That your children's childhood goes by in the blink of an eye? That at the end of life what seems to matter most are the strength and authenticity of your relationships? I'm grateful that people have told me this, because in the moment it is easy to be fooled by the urgency of things. As the parent of a young kid it *feels* very important to focus on practicalities, ensuring there is enough money coming in the door and our family's daily life is stable. But my Elders tell me this is not the whole picture, and in retrospect I have often found them to be correct. It takes effort and a certain amount of faith to break away from my habits of busyness in order to value, for instance, unstructured time spent together. If I don't make that effort, I may have regrets later on.

In the same way, if you never really ponder what space and time are, you will probably survive just fine, but you may get to the end of your life and look back with regret. If you assume you understand how the world works at this basic level, what might you miss? What is fate? What is destiny? Were there opportunities that you couldn't see because you didn't know what to look for? Is the world illusory, as Hindu philosophy claims? What about the cause and effect of karma? Do we understand synchronicity? Before any of these questions are written off, it's important to recognize that they are very difficult problems that have been around for a long time. I think it is unlikely that they have a simple answer, and denial that they exist is too simple an answer for my taste. More likely, I think, is that even with the advanced state of our knowledge of fundamental physics we have more to learn about space and time. What we will learn in the future almost certainly seems ridiculous within our framework today, just as quantum tunneling or superconductivity would have seemed ridiculous to a physicist 150 years ago. For these reasons, though it may at first seem abstract, we can benefit in practical ways from ideas that stretch our understanding of space and time.

In this chapter I will describe my own work in more detail. It is not yet clear whether or to what extent it is correct. It appears to be consistent with modern experiments, and it is based off years of investigation inspired by a core curiosity that awoke in me early. In *Living in Flow* I described a process I call *retroactive event determination*. Our world is primarily about experiences, but the physical situations we experience "fall into place" only when we experience them. Because of retroactive event determination, the history of the world around us is flexible, rendered on-demand right at the moment we experience it. Here I describe this phenomenon in more depth based on the physics of holograms. The world begins to seem like virtual reality.

It should be noted that I present the concept of the holographic multiverse in a way that, to my knowledge, is new and different from previous formulations of quantum mechanics. It is not a theory that has yet been accepted in mainstream science, though this is the focus of my efforts.[2]

It starts with the work of David Bohm.

Bohm and the Implicate Order

Physicist David Bohm was on the leading edge of this way of thinking about the world with his book *Wholeness and the Implicate Order*. He recognized the importance of light's holographic properties and what they showed us about space and time. His conception was a universe consisting of two related "realms," the explicate and the implicate order.

The explicate order is the usual physical world you are familiar with. It contains everything that is real and measurable. Yet, he proposed, there is a hidden order. In the implicate realm there are patterns of information that give rise to the external world we see. The various branches of the tree of possibilities exist in the implicate order. The mysteries of the double-slit experiment, in which particles can interfere with themselves and get canceled out, reflect calculations going on in this implicate realm.

This worldview forms the basis of the research I have done into space and time. As Bohm understood, the relationship between these two realms is the Fourier transform. Together they interact much in the way a hologram works. The image we see in the hologram is like the explicate order. It seems real to us in its detail and relief. Yet it is the result of information encoded into the holographic film. When we tilt our perspective on the film, we see a different image, a different explicate order. Bohm explains, "In each region of space, the order of a whole illuminated structure is 'enfolded' and 'carried' in the movement of light."[3]

The most profound aspect of this is that the physical world we see is dependent on the hidden world we don't see. Bohm says, "In the formulation of the laws of physics, primary relevance is to be given to the implicate order, while the explicate order is to have a secondary kind of significance."[4] Because of this, and because of the one-to-all and all-to-one nature of the relationship between these realms, there is a hidden connectedness in the fundamental structure of nature.* Bohm understood too that this holographic aspect of the universe had to extend beyond the familiar model of a hologram in space to a hologram in time as well. He understood that time is the way we measure change, so that if we include time in our hologram, then we've lost the ability for the hologram to change in time.

To understand this, think about a novel you can remember reading. The storyline of the book takes place over time. As you read the book, the characters change. Each point in the book corresponds to a different point in time. But the book as-a-whole contains the entire story. The book is timeless. We have encoded the history of the characters into the symbols printed on the pages of the book. The plot evolves as you turn its pages, but the book itself does not change.

In the same way, a hologram that describes both space and time contains the entire story of the universe, or rather, all the possible stories. This is the very *holographic multiverse* we've been exploring. It contains

---

* You may be familiar with the idea of entangled particles, which indicate a connectedness that transcends space. The connectedness put forth here and in Bohm's work seems to be of a different nature. It may be that the two are connected, but this has not yet been clearly demonstrated.

a description of how the entire universe changes in time, yet, like the book, it does not *itself* change in time. It contains time within it, just as a book contains the history of the characters within it, and just as holographic film contains an image hidden within it. Bohm says, "Reality as a whole ... is not to be regarded as conditioned ... (because) the very term 'reality as a whole' implies that it contains all factors that could condition it and on which it could depend."[5] He continues,

> The movement of ... light implicitly contains a vast range of distinctions of order and measure, appropriate to the whole illuminated structure. Indeed, in principle, this structure extends over the whole universe and over the whole past, with implications for the whole future ... A total order is contained, in some implicit sense, in each region of space and time.[6]

Just as with all our examples of wholeness, the total is greater than the sum of the parts. Bohm emphasizes that "there is no way ultimately to reduce the implicate order to a ... type of explicate order."[7] The hidden realm that organizes what we see is of a fundamentally different nature, connected in wholeness even while the world we see appears full of distinction and separation. Some aspects of the cosmos we cannot understand through breaking it into pieces.

Are We in a Simulation?

In 2003 Nick Bostrom captured the public and academic imagination with an article titled "Are You Living in a Computer Simulation?" His argument was based on the exponential growth rate of technology. Given the major advances over the past decades, it is only a matter of time before we are so advanced in our computing power that we can create realistic-seeming simulations of ourselves. In fact, we could create many such simulations!

What would it feel like to be in such a simulation? It seems possible that a good simulation would not feel like it was simulated. So, the argument goes, given the likelihood that we develop the capacity to do this, and that if we have the capacity to create simulations we'd probably create many of them, and that they would feel altogether real, it is statistically probable that we ourselves are among the simulated beings. We feel real, but we are not.

I don't find Bostrom's way of thinking about what it means to be in a simulation very compelling. In his view, a simulation feels real but is "really" happening on a hidden computer somewhere. He doesn't appear to define what *real* is, seeming to rely instead on a vague comparison to our daily life. He assumes there is a basement-level reality where the stars and everything else actually exist. He says, "Then it could be the case that the vast majority of minds like ours do not belong to the original race but rather to people simulated by the advanced descendants of an original race."[8] There is an original, real world that is governed by real physics, and everything is measured in relation to this.

> The physics in the universe where the computer is situated that is running the simulation may or may not resemble the physics of the world that we observe. While the world we see is in some sense "real," it is not located at the fundamental level of reality ... Reality may thus contain many levels. Even if it is necessary for the hierarchy to bottom out at some stage—the metaphysical status of this claim is somewhat obscure—there may be room for a large number of levels of reality, and the number could be increasing over time.[9]

He acknowledges that we don't really know what is meant by "real" at the bottom level of the hierarchy of simulated realities. Supposedly a computer in the "real," physical, biological world supports the existence of simulated worlds. There exist both "biologically created minds," the real people, and "simulated minds," the simulated people. The computer running the simulation exists at a real location in a real space and time, but the people in the simulation experience a space and time that doesn't really exist. In *The Matrix*, the real world is inhabited by machines that have taken over the world, and the illusory world where people think they live is completely false.

But in truth, if we don't even understand what space and time are, how can we go about distinguishing between "real people" and "simulated people"? How can we claim a difference between a "bottom" level of reality and the virtual ones derived from it? If I am in a dream, I shouldn't use my perception of the space around me as the definition of reality. Bostrom assumes the existence of a real world, to be distinguished from a virtual world. All of his virtual world depends upon the existence of a real world, whose "metaphysical status ... is somewhat obscure." But if we don't understand space and time, we are not in a position to assume there is such a bottom level of reality, a world of stars and planets that we would somehow identify as the real, physical world.

In one of the most influential papers of the twentieth century, physicists Albert Einstein, Boris Podolsky, and Nathan Rosen defined reality in a procedural way, claiming, "If, without in any way disturbing a system, we can predict with certainty ... the value of a physical quantity, then there exists an element of physical reality corresponding to this physical quantity."[10] This doesn't make any assumptions about what space *is*, only about how we experience it. Bostrom's assumption that there is a bottom layer of reality doesn't fit within their definition.

I feel Bostrom's approach misses the point. When we think of a simulation, we shouldn't be focused on "what is really happening" or "who created the simulation." We should instead think about our lives, whatever they are, as "real." The urgent question is, "What we can learn from life?" If our questioning of reality is motivated by a desire to escape it, then isn't it worth asking, "Why do we wish life to be a fiction?" Maybe it is because we want to avoid pain. Maybe it is because we want to let go of our worries. When we are immersed in the beauty of life—enjoying time with friends, making love, watching a sunset, traveling on vacation—we do not seek alternate explanations of reality. We define life as the experience itself, fulfilling and satisfying in its own right, without needing assurance that indeed something physical is there to justify our experience.

Is there another way to resolve pain and worry than by trying to escape from it? Are we questioning reality in order to avoid unpleasant feelings? We might think, "The world isn't actually real, so losing my job doesn't really matter." This is a form of "spiritual bypass," where we avoid the difficult spiritual work on ourselves by becoming enthralled with convenient insights.

I think the concept that the world is a simulation is *mathematically correct*, but I don't agree with the ontology or meaning that many people associate with this idea. The word *simulation* itself conjures up a sense of carelessness and purposelessness. I would instead say that the world is holographic. For me this enhances our understanding and value of life, encouraging us to double down rather than give up on the world. In *Living in Flow* I based my framework of the cosmos not on physical things but on experience itself. Simulation, hologram, or otherwise, having a child is an incredible experience. *Experience is what makes it real.*

The mathematics of holograms seems to accurately describe the physical world. Consider a hologram of a face. The world we experience seems to be a simulation in the same way that the image of the face in the hologram simulates the face. The *image of the face* is really there, but it is only a pattern of light, only a pattern of information. Similarly if the world itself is holographic, everyday objects are really there, but they too are just patterns of information. What makes something real is our experience of it.

We can now finally discuss the possibility that the world is a hologram in both space and time.

The Mathematics of the Holographic Multiverse

The study of holograms is the study of patterns, and the language of patterns is *waves*, as we saw in Figures 13.1 through 13.6. Not coincidentally this is the same language that describes everyday matter through the science of quantum mechanics.

From 2017 through 2020 I researched the connection between the Fourier transform and quantum mechanics. Inspired by the fact that this mathematics appears again and again throughout the calculations of quantum mechanics, I wondered whether we might have the relationship backward. Instead of pattern space being a useful tool for calculations in quantum mechanics, maybe quantum mechanics was the natural outgrowth of pattern space. In other words, maybe pattern space and the Fourier transform are the basis of reality, and quantum mechanics is the natural set of laws that come from this.

I found indeed that the basic equations in quantum mechanics describing the way a particle moves from one point to another, called *wavefunction propagation*, were simplified when seen this way. I had also learned that exactly the same mathematics appears in the field of optics and diffraction, the study of how light travels. It was therefore natural to make the leap that they are part of one single, more comprehensive theory, which is not a new idea.

But this line of thinking allowed for a couple totally new ideas as well. First, it shows us that physical objects can be represented in pattern space, in which they are described as-a-whole. I see this as the realization of Bohm's vision of the implicate order. Second, it gives us a novel understanding of time. Just as the Fourier transform is equally comfortable processing a photographic image on a piece of film or a musical waveform in time, the Fourier transform approach to quantum mechanics allows one to treat space and time equally. This means that both space *and time* can be described as-a-whole. The notion of timelines rather than moments, and a branching multiverse tree, both emerge from taking pattern space seriously.

An example of this was shown in Figures 13.7 through 13.12, where we capture an entire image composed of many buildings, each having complex structure, with a single frequency domain (or pattern space) graph. Figure 13.12 is a pattern that describes Figure 13.11 *as-a-whole*. Currently, since quantum mechanics is thought of as reductionist,* we start from the picture that everything is made up of individual, microscopic particles. Then

---

* Mostly, except for the phenomenon of entanglement.

we assign a wavefunction to each of the trillions and trillions of particles. A very complicated picture emerges, and the quantum effects of each individual particle tend to wash out, or "decohere." Thus quantum mechanics becomes unimportant when describing the macroscopic world.*

But the metaphor of a hologram entices a different line of thinking. Just one pattern space describes the whole city image. We don't need to think about the individual particles. Instead we focus on the shapes of the macroscopic items as-a-whole. It may be that we can start thinking about quantum mechanics very differently, not as a theory only of the microscopic but of the macroscopic as well.

As was pointed out, the Fourier transform approach doesn't only treat space as-a-whole, as in a hologram. It treats time as-a-whole also. A single pattern space describes the entire map of light traveling through space. So our choices influence entire histories. Timelines split off from each other in a mind-boggling array of diverging branches, a tree of possibilities. History is not fixed but flexible. This is the holographic multiverse, a collection of many different versions of the universe.

So how does this really work? Remember from the preface and first two chapters that a hologram is different from a regular photograph because when we illuminate a holographic film with a laser, an illusion appears that seems three-dimensional. The imagery changes as you shift your perspective. Therefore, the thing you are looking at appears realistic, as illustrated in Figures 5.4 through 5.6. How is this possible?

The crucial thing to recognize is that in a hologram *space* has two distinct meanings. One "space" describes the *coordinates* where a given building appears. This is what you normally think of as space: the location of objects. But what is the hologram really? It is a piece of film covered with vague, wavy lines called *interference patterns*, as shown in Figure 5.2. These interference patterns are the other "space." The building is an illusion that is only there when the laser is shining, while the

---

* Quantum field theory improves on this by talking about fields instead of individual particles. The approach I have taken is different from, though probably related to, that taken in quantum field theory.

wavy interference patterns are visible on the film whether or not we shine the laser.

Now, your imagination will have to work a little harder here for a moment. The triangle in the figures helps mark the actual position (the parameters) of a point on the holographic film, where the real interference patterns lie. Think of it like a sticker we placed on the surface of the holographic film. As we tilt the film with our hand, of course the sticker moves wherever the film moves. But the illlusions of buildings (the coordinates) *move* relative to the sticker! It's clear now that the numbers describing the position of the triangle sticker are different from the numbers describing the positions of the buildings. The triangle sticker is in the middle, say at the 1,000th pixel from the left and the 500th pixel from the top, and it stays in that place on the film no matter how we tilt it. The buildings, however, move around. The building that starts at pixel (1000, 500) doesn't stay there when we tilt the film.

It is crucial to realize that these are different things. The building's coordinates move relative to the film's parameters. This same idea is the basis for the world as a hologram. Everything you experience in the world—the people you meet, the food and drink you consume, the Sun you see rise and set—results from physical interactions you have with the world. These interactions are described by their coordinates. They are holographic images that move through physical space like the holographic buildings move across the film. Your daily physical experiences are Bohm's explicate order, implying that underneath there is an implicate order, an unseen piece of film whose patterns of information encode what we see in the "real" world.

But we can't see the film! Where is it? It is wrong to think that there is a physical layer that encodes our reality like the holographic film does. It is not a membrane sitting at the edge of the universe with all of history written on it, or stars and planets projecting through it. The hologram is *pattern space itself*, the implicate order, the "dark side of the Moon" that is always present yet perpetually hidden from sight. Just as light passing through a pinhole displays a hidden structure in every pixel, the possible histories of the universe are present in every mote of matter. The holographic film—pattern space—is available everywhere, present in every corner of the universe.

A map of things past, present, and future is written as interference patterns in the frequency domain all around us. This should inspire a deep sense of awe, shifting our understanding of where we lie in the cosmos. We are not isolated on a planet spinning in space a long way from the galactic center. We are intimately connected to the source of patterns that dictate the unfolding of the entire cosmos. The expression of history as a map in pattern space leads to a concept, retroactive event determination, that has dramatic consequences for our understanding of choice. Indeed, this concept is an extension of what John Wheeler called the "delayed choice effect," which we'll explore shortly. Choices are not only made in the moment, but they retroactively update the past. This phenomenon is well documented and uncontroversial in the behavior of microscopic particles, but when extended to the domain of everyday occurrences, it becomes quite controversial. Let's dive into what it means to be in a hologram of time!

History as a Filmstrip

"Choice" is about how we respond to the objects in space around us. Most of us take for granted that we live in space. We are used to navigating the physical space in our home, our work, and our local geography. We are intimately familiar with space, so much so that we hardly realize our reliance on it.

But what is space? It's a map, a means of telling one place from another. It's a means of distinguishing choices. I can sit in either this chair or that chair. They are distinguished primarily by where they are in the room.

Thinking of space as a map helps us imagine other kinds of space as well. Pattern space (aka frequency space or the frequency domain) is the twin cousin of the physical space you are already familiar with. I used the metaphor of the dark side of the Moon because pattern space is always present but never visible to us. It, too, is a map. Pattern space allows us to distinguish high-frequency patterns from low-frequency patterns. A high-pitched squeal happens at a different "place" in pattern space than a low-pitched moan.

Now we will apply these ideas not only to space (like we've seen in digital photography) or time (like the audio MP3) but to space and time together. Just as there are two notions of space in a hologram—the positions of the markings on the film and the positions where the illusory image appears—in a holographic universe, there are two notions of time as well. This new way of viewing time will allow our choices at certain moments to select meaningful paths through history.

Earlier we could shift our perspective on a hologram in space and see different views of the city. Now when we shift our perspective on a hologram in time, our different views are like different points in history. History is a filmstrip of possible scenes programmed all-at-once into pattern space. When you say, "I bumped into my colleague in the hallway at lunch," you are describing an event in space and in time. But if the world is holographic, the only things that actually exist are those you actually experience. In other words, in between your observations, physics has nothing concrete to say.

Discovering a New Field of Study

Story contributed by Judith Glick-Smith

My PhD studies of transformation and decision-making led me to the idea of "flow." In a random conversation with my brother-in-law, a battalion chief from a northern Virginia fire department, I described the characteristics of flow. This prompted him to tell me about the line-of-duty death of a firefighter in his department. My brother-in-law, the incident commander on scene, made the call not to rescue the firefighter. I realized that all the characteristics of flow were present in my brother-in-law's decision. His story initiated my quest to investigate the flow experiences of firefighters. Daily unexpected synchronicities continue to validate my choice to use fire and EMS departments as a model for flow-based decision making and flow-based leadership.

It's as if the universe is on autopilot whenever we are not around to observe it, traveling along many potential branches but not choosing an individual one. Our interactions are like fence posts that determine the ultimate path of the fence but leave it freedom in between. Between these posts is only a vague description of what *could* be happening. There is no definite reality.

Recall the metaphor in chapter 5 of driving to work in an autonomous vehicle. Just as the entire road plan to get from home to work is decided in advance by the GPS device, wherever you go in life the entire journey is defined as-a-whole as possibilities in pattern space. You might *think* you are making an inconsequential decision to get coffee in that moment, but you are actually choosing between distinct outcomes available to you in the future. There are timelines in which the detour to get coffee leaves you stuck behind a car accident that blocks the road, making you late for a meeting and getting fired. When you make choices in the holographic world, all the possible consequences of your choice are already programmed into the hologram.

It is very controversial to suggest that this may be more than a metaphor. It is widely believed that quantum mechanics doesn't apply to macroscopic objects, a perspective I disagree with. But to make our point without controversy, let's return to studying light.

In *Living in Flow* I used the example of light traveling through space to argue that all the possibilities must exist in parallel, what is called *superposition*. It is worth examining that situation again, through the lens of the holographic description.

Light traveling from Sun to Earth takes eight minutes to make the trip. Yet that entire path of travel is encoded all-at-once. Pattern space is like a map that shows the whole route. In *Living in Flow* I explained that the separation between events is calculated as the time between them subtracted from the distance. This comes from special relativity. For light, these two numbers are always the same, so the separation between the start and end of its travel is always zero, or a null interval. Chemist Gilbert Lewis describes this as "virtual contact" between the Sun and Earth. The Sun and the Earth … *in contact*?! Even in the physics of a century ago, we could see that the path of light traveling through space had to be addressed as-a-whole. The beginning could not be separated from the end.

Thinking of this in terms of pattern space again, the ending point—landing on Earth—is part of the whole map, which includes the starting point as well. Light can't even leave the Sun without a map that describes its journey to Earth. But then, does it have a choice? Is it destined to reach Earth rather than the Moon or Mars? What if a satellite intercepts it before hitting Earth?

These are mind-boggling questions that this new theory of time has a decent answer for: there must be a plenitude of maps (in pattern space) describing the path light *could* travel, each one describing a unique end point and allowing for all the available choices. This is just like the branching tree of quantum mechanics. Pattern space describes many potential journeys over all history, without any preference for what we experience as "Now." It is a hologram in time.

Retroactive Event Determination

We have come back to *choice*, a sort of "timeless choice." The photon traveling from Sun to Earth in the last section is not destined to hit a specific location on Earth, or elsewhere, because there are many different maps in pattern space that could guide its path.

Let's look at two of these possibilities. We set up a satellite in orbit around the Earth with an adjustable mirror. (See Figure 14.1.) A choice can be made by a human being whether to insert the mirror in the light's path or not. If they insert the mirror in the path, the light from the Sun gets deflected toward Chicago. Otherwise, the light is unaffected and continues on toward Houston.

When does this choice happen? The answer is not so obvious. According to the person moving the mirror, the choice must be made shortly before the eight-minute mark, for that's about when the light would finish the trip from the Sun and arrive at the mirror. But pattern space describes the *whole journey* from Sun to whatever destination.

Let's say the map in pattern space describes a trip from the Sun to Houston. This map might look like Figure 14.2. Since pattern space doesn't measure time at all, this map has always existed in pattern space to guide the journey of the lonely photon, somewhat like DNA exists at our birth to guide our growth into adulthood. Its path from beginning to end appears to be predetermined.

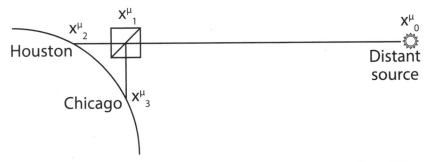

FIGURE 14.1. Light travels from a star far away on the right to a satellite orbiting Earth. When the light has already traveled halfway, a person makes a decision whether to allow the light to pass through the mirror toward Houston or to deflect it toward Chicago.

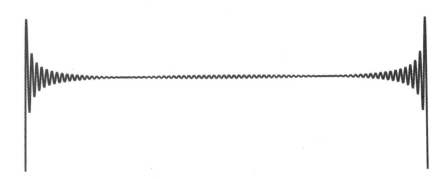

FIGURE 14.2. When drawing light traveling from a star to Houston in the frequency domain, this model predicts a unique pattern of variation based upon the whole path from beginning to end. This type of pattern shows up in electrical engineering and in quantum mechanics when a signal starts or stops abruptly.

But now the choice to put the mirror in the way happens *after* the light leaves the Sun, once the light is already traveling according to information in the map. Because the mirror is inserted, the light reaches Chicago instead of Houston. But there must also be a map for this, a different map, from the beginning. This might look like Figure 14.3. Note that Figure 14.2 is different from Figure 14.3.* The critical question is this: If pattern space does not have any time associated with it, *when (in pattern space) did this decision to move the mirror happen?* We cannot answer that question. Pattern space cannot change at, say, 3:00 p.m. on a Wednesday, because in pattern space there is no such thing as 3:00 p.m. and there is no such thing as Wednesday!

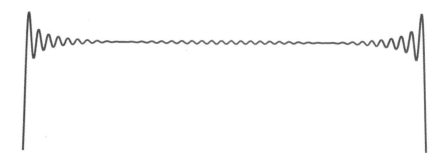

FIGURE 14.3. If the same light is abruptly detected at Chicago instead of Houston (see Figure 14.1), a uniquely different pattern of variation occurs in the whole frequency domain. Since we're discussing not only space but time too, both the "Chicago pattern" and the "Houston pattern" must have always been true as-a-whole from the very beginning of travel.

---

* Although it is not clear in my research that these graphs are the *correct* drawings of pattern space for the light, it is clear that there must be *some* pattern, and that the pattern is different for each situation as-a-whole.

Not only does this mean that every possible outcome must be represented from the very beginning and on into the future by its own map in pattern space, it also means that only the interactions themselves are real. Expanding our scope again with the controversial idea that this applies to macroscopic objects like cars and clouds and you and me, life is like a ladder, and choices that happen later in time affect the entire history in between the previous rung and the present rung. This reality is not about *things* but the interactions *between* things, the moments when things come together. Each interaction defines a new branch on the tree of possibilities. Things change at every splitting of the multiverse. Yet along a single branch, in between choices, history is defined as-a-whole and our choices play out. Destiny unfolds.

Let's picture a baseball flying through the air. This theory describes the ball's inevitable journey from its beginning point to its end point. Because quantum mechanics is the study of what the world is doing when *you* are not watching, the ball proceeds undisturbed unless *you* interact with it. The arc of the ball's travel is given as-a-whole, and the "motion" of the ball is just a set of points in space and time with a certain mathematical relationship. We have to let go of the sense that there is a master clock. It is you who brings the concept of "Now." It is you who experiences the ball as moving through space. Each observation you make of the ball affixes it at a particular point along its history. But from a more fundamental view, not the one limited by you the observer, motion itself is more abstract. It is a way of viewing the images inside the hologram from a particular perspective.

Each day you wake up and find yourself at a particular place and time in the hologram, with a unique view of circumstances. You are tapping into the great flow of the infinite, experiencing for yourself this one precious moment, yet seeing that it is connected by a thread to all of history, both actual and possible.

To picture this timeless, spaceless, holistic path, imagine the touring schedule for a musician. I love singer-songwriter David Wilcox and recently looked to see if he would be performing anytime soon. On his website I found a set of tour dates. These were the locations he would be at on each calendar day. The tour schedule is like pattern space. It contains all the possible matches of where and when one can find David Wilcox. If I want to see him in San Francisco, I need to show up on March 5 at 8 pm. Or if I know that I am free only on March 10, then I need to fly to Portland, Oregon. A path is simply a relationship between place and time. But this analogy shows us that the path itself—the tour schedule—is sitting as-a-whole on his website. I can think of David Wilcox existing at all the places and times listed on the tour schedule. In between those dates, I can assume he is traveling on a plane or bus, but I actually know nothing of the sort for certain. I only know the moments when he is on stage and I am listening to him perform.

Like rungs of a ladder or posts in a wire fence, events in a holographic universe occur at intervals, with undefined spaces between them. Each experience is a single interaction or perception, capturing the world right here and right now. But in between the fence posts or ladder rungs is spaciousness, or *spacelessness*, an undetermined, amorphous tree of possibilities, a shifting ocean of potentialities.

The photon doesn't travel smoothly from the Sun to Earth. Rather, it is witnessed at Earth, and only then do we infer that it *had* traveled here. Events don't actually "happen" when they happen. They become definite only after the fact, when a measurement requires them to commit to one history or the other. This is *retroactive event determination*, a phrase that captures the idea that you are not changing the past from what it already was but rather selecting from various possibilities that might have been.

You can picture retroactive event determination like a "get" email system, in which your inbox only loads emails when you log in, rather than downloading them behind the scenes. Your physical reality is not continuously delivered to the inbox on your desktop computer. Rather, it only downloads when you log on to the internet. At that time, however, you get all your emails as if they were delivered smoothly over time. Each email has a date stamp corresponding to when it was (supposedly) sent. Did the email really "happen" at the moment on its time stamp or at the moment when it entered your inbox? The world moves in jumps and starts, becoming actualized only when you check on it. Yet when you check, life appears to have been happening steadily all along. A photon doesn't have a definite path, except when you investigate to find whether it did.

This is an idea understood early on by physicist John Wheeler in the middle of the last century. Wheeler was Richard Feynman's teacher and advisor, a great inspiration for many researchers in the field. He illustrated the retroactive nature of quantum mechanics with a thought experiment called the "delayed choice experiment." Physicist Kip Thorne describes the experiment and Wheeler's view of it:

> The central facet of an (ideal) quantum measurement, John maintained, is the "collapse of uncertainty into certainty," as embodied in the collapse of the wave function. He probed this collapse conceptually with his "delayed choice experiment," a thought experiment in which the experimenter's choice of what to measure can be regarded as influencing the past history of the measured system … The choice of measurement … could then be regarded as reaching into the past and making definite which path(s) the photon has followed: one or both.
>
> Several years after John conceived this thought experiment, William Wickes, Carroll Alley, and Oleg Jakubowicz actually carried it out at the University of Maryland, getting precisely the result that John knew they would.
>
> This thought experiment led John to speculate that the universe might be "a self-excited circuit," a system whose existence and history are determined by measurements, many of them made long after it came into existence …
>
> John's self-excited-circuit idea in turn led him to speculate that information theory is the basis of existence: "Trying to wrap my brain around this idea … I came up with the phrase 'it from bit.' The universe and all that it contains ('it') may arise from the myriad yes-no choices of measurement (the 'bits') [that occur during the life of the universe]." As crazy as this may sound, many quantum-information scientists think it respectable. In John's famous words, it just might be "crazy enough to be right."[11]

In a computer, a *bit* is a discrete thing. The word is a contraction of the words *binary digit*, where *binary* means "having two parts." A bit has a value of 0 or 1, and nothing in between. There are no allowed values between 0 and 1, and there are no gradual transitions between one bit and the next. It's true that the memory of your computer contains many electronic circuit elements to transport electric fields—wires and such—but there is only one thing that really matters: the electric field in each storage element that represents a 0 or a 1. In using your computer, you are only aware of the activation or deactivation of bits. These tell you whether your rideshare will meet you on 10th Street or 11th Street, or whether your paycheck will be $100 or $1,000. Like with rungs of a ladder, you don't care what happens in between the bits or how they travel through the computer, only what information they represent in the end.

This is how we describe the photon traveling from Sun to Earth as well. In between Earth and Sun there are only irrelevant details. In fact, there is no definite reality to these details. They are undetermined and fall into place only once the final destination of the photon is determined. So it is not surprising that we can reach back eight minutes in time and "modify" the past. We aren't in fact doing so, for the past did not exist yet.

If we take Wheeler's phrase seriously, as many physicists do, we conclude that in between emission and absorption of the photon there is no fixed history, no actual reality.

The Software of the Multiverse

Wheeler's delayed choice thought experiment illustrates that the cosmos is made of information, and this information doesn't tell you where something *is*, but rather where you would find it *if you were to look*.

This is a crucial difference from our current way of thinking, one that should inspire a moment of awe. An object does not have an "actual history," only the history that falls into place at the moment you interact with it. History is not set in stone (or "set in time").

Recall Kato's job search from chapter 5. He accepted an imperfect job opportunity, wishing he could find something closer to home. Then at the last minute he received a call that the position he really wanted had suddenly opened up. He had no information about how the situation was evolving behind the scenes, only that the final result was a job opening for him. Histories are flexible. What "has already happened" is not yet decided, so long as you have no data about it. The certainty of life events is delayed until you make a choice.

You would not be surprised to find this flexibility as part of the software in a flight simulator or virtual reality training program. For instance, a flight simulator would only create bad weather if your instrument navigation skills were being tested. The circumstances needed for a certain educational experience would not be created until it was clear they were needed. And when the bad weather happens, it is conjured out of thin air. The storm doesn't have a real history to it; it has been fabricated to match the need.

Remember the metaphor of a fence. Your interactions with things are like the fence posts that you can call real. In between the fence posts are just wires of imagination or linked chains of possibility where options are developing but nothing is certain. The hologram contains all the possible realities. Your particular view of the hologram provides you the specific experiences that you call real life. You have navigated to this view via the choices you made to get here, a Choose Your Own Adventure novel. The world is rendered for you on a need-to-know basis only.

This is known as *contextuality*. In a contextual system, the interpretation of one piece of data depends on another piece. Consider the phrase "The old train ..." This is the wire in between the fence posts, the undetermined past. Why is it undetermined? Because it is not yet clear what the meaning of "train" is. Let's say the end of the sentence is "... left the station." This is a fence post that makes it clear that "train" refers to a locomotive, and the locomotive was manufactured a long time ago ("old").

However, a different end of the sentence could be "... the young," which changes the meaning of the sentence to a statement about education rather than railroads. Did you *change* the past? No. It existed in a superposition of possibilities until the final words were uttered. You can't claim to have changed something that didn't already have a definite form.

This example of retroactive determination of meaning is called *local ambiguity*. It is an apt analogy for how a holographic world would allow the paths of objects to fall into place based upon the choice of action you make at the *end* of the journey.

In *Living in Flow* I described the *optimistic synchronization* process, which is a software technique used to overcome hardware limitations. Video game designers use optimistic synchronization to allow the game players to create a seamless virtual world. Each player has their own rendition of the virtual world, but those separate worlds are experienced as a single, integrated world by making sure they remain mutually consistent with each other.

Retroactive event determination is the way in which our world is like a video game. This is the virtual-like, illusory aspect of reality. We experience our own version of the world, rendered on demand, and the mathematics of quantum theory ensures that our worlds remain consistent. We do not experience an objective reality but a relational one, defined by our interactions. Our lives are an ever-growing conspiracy of correlations.

If you and I meet for coffee, that event becomes a fence post in my fence. From my perspective, our interaction intertwines our histories together. You are no longer free to have been anywhere else, for you were at coffee with me. Yet from another person's perspective, both of us are part of the gap between their fence posts. From their perspective, there are some branches of the multiverse on which we are together having coffee, and others on which we are each doing our own thing. We must always think about whose perspective we are describing the world from, and then think critically about what facts are really known from that perspective. Facts such as "who met whom" are relative facts, not yet determined to those on the outside.

This is our world as a hologram. It is flexible and responsive. It is whole, spaceless, and timeless. It provides us only the barest glimpse of this enormous cosmos, yet that glimpse is all there is to see. We are the center of our world, its primary beneficiary, and simultaneously just one portal on the magisterial halls of blackness we call outer space.

Inspired by physicist Niels Bohr, John Wheeler espoused a methodology called *radical conservatism*. "Base your research on well-established physical laws (be conservative), but push them into the most extreme conceivable domains (be radical)."[12] If we push the limits of what a holographic formulation of quantum mechanics would tell us, we find ourselves with a radically new picture of time and the importance of relationships.

Access to truth is shared equitably among all participants in this virtual-like world. The template that guides each proton or person or planet along its lonely course through space and time is available to each and every viewer living in each and every habitable planet in each and every corner of each and every galaxy of the universe. We are *not* separated. We are more connected than we can possibly imagine. Every little change we make is felt across the entirety of creation. If we understand the wholeness of which we are a part, we will be perpetually overcome with awe and wonder.

15

MAKING THE LEAP
TO WHOLENESS

The breeze at dawn has secrets to tell you.

Don't go back to sleep.

You must ask for what you really want.

Don't go back to sleep.

People are going back and forth across the doorsill

where the two worlds touch.

The door is round and open.

Don't go back to sleep.

JELLALUDIN RUMI[1]

The Leap Is a Feeling

We have spent the passages of this book analyzing space and time, emotions and thoughts. Yet there remains something uncapturable.

Choices are scary. To grow and change we need courage to do things differently. When we try new ways of being, the momentum of past habits yells at us to do it the old way. Nobody takes our hand and says, "Yes, you may be more confident now" or, "Don't worry, it will all be OK in the end." In the final moment, we must make a *leap* from the old way of doing things to the new way.

We are no longer stuck in the old filters and ways of feeling. We have practiced moving through all the inner rooms in feeling-space. Finally, now, we choose the new way we want to feel. We jump off the edge.

Maybe it's a new conversation with our parents or friends. Maybe it's a business we want to create. Maybe it's learning a new skill. Maybe it's bringing up a difficult topic at work. Whatever it is, it is a choice that makes us more whole.

Wholeness is not something we are given when we earn enough credits or do enough inner reflection. Wholeness is not acquired, awarded, or permitted. Wholeness is chosen. Choosing wholeness doesn't require permission or justification. No matter how much we accomplish in our lives, we will not feel whole until we *choose* to.

Similar to the image of the city in Figure 13.11, our whole self is made of all the components that make us who we are. We are not made of bits and pieces but of patterns. Each component is essential and can make itself known in every corner of our lives. If we carry unhealed resentment from a past romantic relationship, that resentment may show up with our friends, with our family, and with people at work.

To *feel* whole requires a leap. No matter how much preparation we do or how talented we are, we still get confronted with the same familiar situations that make us want to repeat old channels of behavior. We are never immune from feeling embarrassed or free of feeling resentment. We simply get better at finding our way to other emotions even when the habitual ones are clamoring for our attention. We gather the strength to discontinuously leave behind our pain and choose something completely different.

In Chapter 11 we learned how to navigate our way through feeling-space to find the feeling we want to have. Now that the new path has become more well worn, we can leap straight to where we want to be. The leap to wholeness requires courage to face any fears that emerge when we react out of habit, so that we have space to experiment with more authentic responses.

The leap is a feeling we choose. Over the past few months my family has been quarantined, just as so many others have, in response to the COVID-19 pandemic. During this time of close quarters, it has been easier than usual to lose patience with each other. Recently I raised my voice at Ellie because she was watching Netflix when I had asked her to stop. "Why are you watching TV? I thought I asked you to read a book or go outside. Turn it off, now!"

Why was I attached? I guess I was feeling fearful and guilty. I was worried it would become her habit, and I felt bad that I didn't have time or energy to start a science project or something with her. She obviously really wanted to watch, and I was reminded of my own love for video games as a kid. My dad had also been critical of me. I had had to protect myself from his criticism by sneaking to the video game console when I thought he wasn't around. Now my own hurt was perpetuating my daughter's hurt.

Even though I realized the pattern, I found it hard to let go of my anger. Later in the evening, as we were sitting down to read Harry Potter together, I was brooding that I had set up a permanent "I feel bad about it" impression, like I had had as a kid with my dad. I recognized many other ways that I had been critical of her as well, a legacy of hurt I had passed on to her from my own upbringing. Would I ever be able to stop being critical of her?

But there was even something underneath that. Was my obsession with feeling bad just a reflection of the very same thing I felt bad about? Not only had I criticized her, I was criticizing myself. In fact, I had criticized her *because* I was criticizing myself. I had found it hard to smile at her, and no wonder! I was being super hard on myself.

I wanted to make things better by explaining myself and repairing some of the damage that may have been done from my emotionally charged words. But that's not what she needed. A simple smile would tell her that I loved her. Saying sorry and explaining would just reinforce the hurt. She would feel my authentic emotions with every word I said. But my smile would be healing for her.

Smiling was my leap to wholeness. Smiling was stepping beyond the *promise* of love and into the *experience* of love. There was nothing more I could say to repair the situation. I simply had to *be* different.

I took a deep breath and closed my eyes. What do *I* want right now? What will bring *me* joy? How can I find a room of happiness and be at ease with myself?

I smiled at her and she smiled back. We opened up the book, and Harry, Hermione, and Ron helped us shed the pain and choose who we wanted to be.

The Timeline of Choices

There is a middle ground between the materialist, reductionist, causal view of the world in which things in the environment simply are as they are, and the New Age view that anything is possible.

In the middle, we live in an ever-growing conspiracy of correlations and connections. Who we meet or what we do in the future is constrained by who we've met or what we've done in the past. Our future has to be consistent with our past. There is no need to question whether we can *change* the facts of the past. There is plenty of flexible opportunity waiting in those events we simply haven't observed yet. Most of history lies in between my fence posts, unfixed and containing mysteries. Everything that is *not in*consistent is still possible.

In the last chapter I laid out the science plainly, as free as I could from interpretation. Much of the information presented there is well accepted and often experimentally verified scientific theory. What has not been yet well accepted is nonetheless developed rigorously in a testable fashion.

Here, though, I will venture into my personal interpretation of the holographic worldview. It is important to understand our paradigms as they shift, because our paradigms are the filter through which we interpret reality and make new choices. Understanding the model of the holographic multiverse, if it is correct, is paradigm-changing, so I feel it is important to speculate about what the new paradigm looks like. We are at a time in history where new understanding is urgent. But realize that any discussion on the nature of personal choices is left, finally, to the opinion of the person making those choices. While science can set the paradigm, we as individuals must decide how we operate within it. What remains is my own personal takeaway.

I think the Holographic Paradigm provides just the right balance of elements from the old paradigms. A great deal more is possible than a reductionist worldview would concede, yet the Holographic Paradigm allows for a realistic acceptance of the consequences of our past choices. The New Age worldview yearns to comfort our grief by imagining that everything can be fixed, anything is possible. We are better off acknowledging our past, accepting our losses, and learning everything we can about ourselves, so that we can see the possibilities that still stretch out in front of us. By having an accurate sense of what is possible, based on both the realistic constraints of our own past and the possibilities waiting to be found there, the most important thing we can gain from the new paradigm is a wiser understanding of the role of choice in our lives.

The choices you make can affect even those events you may assume have already happened, so long as you don't have any contrary information about them. The world is adjustable, and history is flexible. In this sense, everything around you is information. All that is considered real is that which you have information about. This is Wheeler's "it from bit."

The information that guides your experiences is contained in the patterns of pattern space. In pattern space the world is spaceless and timeless, and possibilities overlap on each other. Just as the navigation system on your GPS describes your whole journey at once, or like sound waves in time that have been converted to pitch information,* the interference patterns in pattern space encode all of history at once. In this model, the facts of daily life—whether or not a certain email comes to your inbox, whether or not your flight gets canceled, whether or not the roulette wheel lands on red—are not the fundamental reality. Rather, the complicated patterns that encode these facts are fundamental.

---

* Here I am referring to what a Fourier-transformed file would really be like, not the time-ordered hack used on the MP3 file that we discussed in chapter 13.

We take for granted the solidity with which we see the world. But infants do not have this "object permanence." For them, the entire world consists of what is right before them. They do not yet have the mental tools to construct a rendition of reality to keep track of that which is not actually there.

Maybe the infant sees truly! The physical objects we observe every day do not persist. In this sense, they are an illusion. They only take on a definite form when we interact with them. What does this say about choice?

Imagine we are notified at noon that we had been sent an email with an important decision about a job offer. Yet we don't check our email until 6 p.m. At 5:45 p.m., can it be considered "true" that the email and its contents exist in our inbox? Could it be that it still hasn't been "decided" which history we will experience? Is it reasonable to say that the contents we ultimately find in that email sent at noon are dependent upon our choices *afterward*?

Let's say that at 3 p.m. we choose to research more about the company and train further in the skills needed for the job. Or, alternately, we spend time crafting our next job application. Could these activities influence the decision communicated in that noon email?

Let's just focus on whether this scenario is logically consistent, putting aside for now whether this is the correct perspective. According to the principle of retroactive event determination, it is not until 6 p.m. that the employer's decision becomes real for us. This is the moment when we check our email and the information must be either one thing or the other. In this sense, selecting an outcome after the fact is perfectly consistent with causality.

In a hologram-like virtual world there could exist what Carl Jung called a synchronicity or "acausal connecting principle," in which the action at 3 p.m. influences the outcome of a decision made at noon, without violating causality. The "causal chain"* only becomes determined when we check our email at 6 p.m., at which time many causal chains can fall into place.

---

* A *causal chain* is a series of events, each caused by the previous. For instance, a pitcher first winds up, then he pitches the baseball, which leads to its getting to home plate, which leads to the batter swinging, which leads to the ball being hit, which leads to its flying into the air.

Let's say we check our email to discover that we got the job offer. This result, and the happy state of the employer who composed the letter earlier in the day, takes shape retroactively at the moment we read the email. Only the interaction between us and our inbox is real (from our perspective). Alternately we may open the email and find that we did not get the job. Thus, those words written in the email, along with the not-so-happy employer who wrote those words, along with everyone who was part of the hiring decision, are part of a history that fell into place in this way. If we were to interview each of those people at this point, they would agree that the decision had been made in the negative.

There could be other versions of the email as well. The hiring decision could have been postponed, or we could have been requested to provide more documents before the employer is ready to make the decision. Each of these sets of circumstances is part of a broad causal chain that can fall into place after the fact when we read the email.

So long as the correlations are all accounted for, this model is completely consistent with the world we actually experience. There is no easy way to poke a hole and determine which view is correct. If we call one of the people on the company's board of directors at 3 p.m. to get the inside scoop, that is just another way of tapping into the information we are seeking. We will always find that the email we read at 6 p.m. will reflect whatever information we get from the board member.

The coordinates at which an actual thing happens—the time at which we check our email and the specific email we find waiting there—are an event encoded into the holographic patterns of pattern space. Even the time stamp when the email was supposedly sent is part of the data we download later, but this is all for show. It *becomes* that only when we check. The time stamp on that email means we *infer* that it was sent some time earlier. We experience the illusion that things were happening all along even though we are only really sure of what happens when we check.

There is a famous Zen koan, or mind puzzle, that reads, "If a tree falls in a forest, but nobody is there to hear it, does it make a sound?" In this new paradigm we have a useful answer to this puzzle. In the absence of you, the observer, being there to listen, the tree remains in a suspended animation of both "fallen" and "not-fallen" possibilities. Who can say which event has *really* happened? There is no such reality. It is only when the observer comes by soon after to find the tree fallen—or not—that they construct a story about what must have happened.

We have then stitched together a persistent image of the world. Clearly the tree seemed to be having whatever its experience was without this observer there to watch it. But because reality is relational, we must choose a perspective if we are to say anything concrete about the world. And every perspective is incomplete. From the observer's point of view, there was no concrete answer to the question "Has the tree fallen?" Thus, the problem presented in the koan is the way the question itself is formed. It is not so much about the sound that the tree makes but about what it means for something to "happen"!

The School of Choices

Though modeling the world as a hologram may seem like an abstract undertaking, it has great relevance for our everyday experience. Too common today is the worldview that we must simply react to a world that is already unfolding. Life in such a paradigm becomes about coping with change after it occurs, believing that "things are as they are" and there is little to be done.

In chapter 3 we heard Joseph Jaworski's alternative view that we can "create the future" as opposed to "reacting to it when we get there." Otto Scharmer holds a similar view. He says, "The essence of leadership is to become aware of our blind spot (our interior conditions or sources) and then to shift the inner place from which we operate as required by the situations we face."[2] Scharmer continues, "What sets us apart as human beings is that we can connect to the emerging future … We can break the patterns of the past and create new patterns at scale."[3]

We can think of the responsive cosmos as a school, not of academics but of the most exciting aspects of life. Each lesson is crafted for the student. For instance, I remember a recent Friday evening when a lecture was happening an hour away from my home. I was striving to expand my professional network, so I left my family at home for the evening and made the lonely trek across the city. In my heart I was sad, because staying home on a Friday to watch a movie with my family sounded so nice. But I was committed to what I was trying to accomplish. Yet when I arrived at the lecture, I found it was small, poorly organized, and not compelling. I quickly wished I was back at home.

There was a lesson here: I could appreciate the joy of being with my family and didn't have to default to the professional obligations every time. I could check in with my own feelings, figure out what I really wanted, and find flow in that way. I hastily made a few connections at the lecture, stayed for the first ten minutes, then hopped back in the car. I arrived home to catch the last half of the movie, and my night was complete.

If the cosmos is a school, it is not a boring one. It might be teaching us to tackle our fear of snowboarding, or follow our love of horses, or gain an appreciation for our best friend's favorite music. It might be teaching us to let go of a lost romance, or to heal from the loss of a loved one and rediscover joy. Situations unfold every day for each of us that, due to meaningful history selection, are personally tailored to us. Whatever interests we have or habits we can't shake, these guide the choices we've made in the past, and thus meaningful history selection customizes the next event after the previous choices.

Even if regular school wasn't a fun experience for you, the learning that synchronicities bring us are always aimed at getting closer to our heart. They peel back the layers of who we are to uncover the part of ourselves that wants to be seen, wants to enjoy life, wants to express ourselves authentically and powerfully, and wants to be connected in meaningful ways. Personally I have learned to trust the process fully. I am always, in the end, amazed at and grateful for the education that synchronicities bring me.

This being the case, it is not my experience that every situation is fair. Suffering is very real, even when we are paying attention to synchronicity, even when we are in flow. In earlier chapters I emphasized that synchronicity as a teacher can bring us difficult experiences. It is useful to think of the situations that happen to us in terms of their context rather than their content. We can experience the suffering of being impoverished, imprisoned, limited, born into an unfortunate family, sick, disabled, frustrated, too short, or too tall, and yet experience just as much learning and healing as people with more fortunate circumstances.

It is not my intention to present this new paradigm of choice as easy or optimistic. The purpose of synchronicity in our lives is not to make life better in the ways that we consciously want it to be. Rather, its purpose is to heal the hidden wounds inside of ourselves. This looks different for each of us, and it's my hope that whatever *your* struggles are, you can use synchronicities in your life to enhance your healing and accelerate your growth.

Living inside a hologram is a shift from resignation to flow. It involves increased focus on what we need to do to get to the next level of challenge, and decreased focus on losing or winning. A world that responds to our choices by bringing appropriate learning experiences is a world that is sympathetic to each of our specific challenges as evolving human beings. Collectively our world is waiting for us to be ready before moving to our next level. What do we, humanity, need to learn in order to get there?

We live in a world of choice. Nothing is more sacred or more powerful than our ability to choose. Yet it is deceptively mundane. Most of our choices seem unimportant. Important life choices do not happen frequently, and we sometimes recognize their importance when they do. We are highly attentive to making careful decisions when it comes to getting married, taking a job, or moving to another city.

But other decisions are less obviously impactful. What am I feeling when I discuss finances with my partner? How do I respond to my child when they do the same difficult behavior over and over? What do I say to myself when I spill the garbage can on the street? Each of these choices is important, for they illuminate our inner landscape. The reaction I have when I spill the garbage will probably be not too different from the reaction I have when I make a mistake in a job interview. The way we are in the small moments affects the way we are when the bigger opportunities arise, so there is no moment too inconsequential to learn from.

We are children of the universe. We are not expected to get things right. The universe is a hot mess of boiling entropy, far beyond our feeble abilities of manipulation. There was a day I asked my father, a retired carpenter, if moving my kitchen pantry wall six inches out would be easy. He replied with a hint of exasperation, "Sky, nothing is easy." The cosmos we live in is messy by design. Life is a steady stream of obstacles, and it is through our confrontation with these obstacles that we are tested. It is through being poked at and prodded that we heal and grow.

The Art of Choices

The difficulty with choices is that we have to make them in the present moment, but each choice is part of an extended timeline that provides important context for us. Steve Jobs famously said, "You can't connect the dots looking forward; you can only connect them looking backwards."[4] When we look back in hindsight, the story of our lives makes more sense than it does in the present. But that story exists all at once as a branch of the multiverse tree, if only we could see the whole branch when we're in the middle of a difficult choice!

So how do we make better choices? How do we find the path to a world with less prejudice, more mutual respect, and better communication? Stuck as we are in the here and now, we can't be completely sure whether the choices we make reflect our deepest values. When we are offered a job, it may not seem clear whether this is the job we want. We can't see the whole picture, so we can't know whether this will lead toward the life we envision for ourselves. But when we look at the whole timeline in retrospect, it becomes clearer how the situation reflected our values. "Yes, this job was a stepping stone toward the next thing," or "No, this job was a distraction from what I really want to achieve."

We have to make choices with the information we have, and so our choices reflect our filters. We decide our actions based upon what we can perceive about our feelings at any given moment. Again and again we experience situations that reflect our choices and therefore reflect our feelings. Gradually we gain clarity, peel back layers of misunderstandings, and make wiser choices.

But we don't need to be perfect. The content of life remains a mess and never gets totally ironed out. Even if we get the desired job, some other part of our life will probably be falling apart at the same time. The question comes back to context. Have we learned what we needed to learn from this process? Has our successful quest to get a job taught us how to be more confident? Sure, maybe we accepted a lower salary than we'd like, or it's only a temporary position, or the work schedule is difficult. The important point is that we have stepped into the next level of personal accomplishment and growth.

In my oral exam for my master's thesis, my three advisors were grilling me not only on my research paper but on my knowledge of the basic physics we were supposed to have learned in the program. One of them realized I couldn't immediately recall the form of the simple harmonic oscillator, a standard example problem that every physicist needs to know.

I sighed in frustration, "I hate the harmonic oscillator," and they looked at me with incredulity. "You have to understand the harmonic oscillator, Sky." Well, I knew that was true, and I did know it well enough to satisfy their questioning as they pushed me on it for a few more minutes. Finally they sent me from the room and made me wait in the hallway while they deliberated. After an excruciatingly long period of time they came to get me and say, "Congratulations, you've done a great job. You've passed." We went down to my advisor's office where they had glasses of champagne waiting to celebrate. I was deeply moved. I felt like I belonged, even though I was no expert on the harmonic oscillator, and they confessed that they just made me wait in the hallway in order to make me feel like it was a serious business!

Perfection isn't a thing in the holographic multiverse. Life is a game that can be played but never won. What we can aim for is to refine our tendency to be reactionary so that we have more choice. We can aim to gain better mastery over our perceptions. The experiences of life make up the illusory content that helps us see our reactions. If I am picking my daughter up from school, most of the content of the experience is

superfluous and habitual. But often a part of the conversation is conse-
quential. Do I remain patient with her when we disagree? Do I manage
my own feelings and model that for her?

Most of those activities we engage in are, in themselves, inconsequential. They simply serve as a backdrop for growth. The city streets we pass
through on our way home, the store we stop at to pick up some milk, the
stoplight we get stuck at for an extra cycle—these are all a backdrop for
the context of the situation. The conversation she and I have and the way
we experience our feelings in the conversation can have far-reaching
implications for both of us, within our relationship and in other areas
of our lives.

I don't know how to make sense of the pain and suffering experienced by people living with oppressive circumstances. Racism and bigotry are systems that have grown and established themselves for a long
time, and it seems inadvisable to apply this idea of "life as a game" to
those situations. I present this model as a tentative first step, a way to
recognize the little choices we make every day and accelerate our journey toward authenticity and freedom.

The content of life provides a stage, but the context of life is the
meaningfulness of our choices. "How am I going to respond to this
person when I see them getting upset?" Or, "What am I going to do
when someone cuts in front of me in line at the store?" Or, "How am I
going to respond when the car in front of me lags and makes me miss
the light?" Each of these represents meaningful moments that jump out
at us and catch us by surprise. These are the moments when our inner
peacefulness is tested. If they trigger some filter in us, then we have work
to do on ourselves.

It pays to remain on the lookout for moments of learning, choice-
points when our filters are active and our perceptions are warped
and twisted. These are the moments when we can "up our game," the
moments that matter because they lead us closer to who we want to be.

As I have gradually come to see my own patterns of negativity with
my wife, and protectiveness with my daughter, and self-pity when I feel
frustrated, and hurt when I get together with my extended family, and

insecurity with my work, I think of myself moving through the levels of the virtual-like world. This happens by peeling back and healing the filters one by one. These filters are the thoughts and feelings I have—the weather—that make me feel separate and end up making me less happy or less satisfied than I could be. Some of the filters I have been able to peel away, others not. They naturally heal once I expose them, because as soon as I understand what's going on it becomes easier to change. I can more easily say, "Oh no, of course I don't need that person's approval. What was I thinking?!" Those healed filters leave room for me to experience greater ease, enjoyment, success, or whatever else I seek that I have been struggling for.

We live in a simulation of choices. But let's be honest with ourselves: wouldn't we really prefer that solutions were handed to us easily? The path seems exceedingly difficult, and maybe this is *on purpose*. By lifting the weights off our *own* shoulders, we become strong enough to handle the freedom that awaits us. Each accomplishment prepares the way for the more authentic way of being whose goodness we could not have accepted before. Through the trial of climate change we may be forced to make changes in how we relate to each other that we would not choose to make otherwise. We may become more compassionate, more understanding, more clear, more honest, and less invasive of each other. Maybe without these changes to who we are personally, society would devolve into violent conflict. Maybe climate change is the path forward into the future we want.

In addressing our problems, there can be complications at every step. In any curriculum, lessons are not always meant to be guidance for us but can often be attempts to cause confusion. By dealing with confusion, we ultimately become clearer on what truth looks like for us. Our sense of reality becomes more refined, simplified down to only the most central perceptions.

When I shaved my head and recklessly set off to another city in search of my future, I caused pain to myself and others. But I'll always remember when one friend, while watching me make these choices, said with a tinge of sadness, "Sky, you're a good person." He saw me struggling to find value in my life and offered me the most basic piece of information he could. As I've matured, this simple statement has become what I think of as a core truth. When I find myself struggling to decide between path A and path B, it is the core truths that I come back to. You're a good person.

Synchronicity is not somehow meant to solve all these problems and make life easier. Sometimes we have to fight against despair to achieve what we value. We don't have to be perfect, but we do have to try really, really hard. To prevail against confusion, we can keep coming back to core truths, whatever those are for us.

Meaningful growth comes in leaps and bounds. Sometimes it is only at the last moment that we find a way to the next ladder rung. Working right up to the deadline is the ultimate prayer, for it is with our commitment and perseverance that we change the world. And in that final effort we can come to *expect synchronicity* to carry us to the next level.

The holographic multiverse tracks only one thing: our choices. It responds to our choices of thought and our choices of feeling, because they ultimately translate into our choices of action. The Holographic Paradigm can be our guide on this quest.

The Leap to Wholeness

How could the world be different in the new paradigm? Our collective behaviors could shift very subtly yet result in far-reaching changes. The more we understand our own filters, the more we will respond with compassion rather than vindictiveness, the more we will be authentic rather than guarded, and the more we will persist in our inner struggle for the world we want to experience.

How do we get there from here? It is an inner struggle. We don't change others—we change ourselves. As our perception changes, gradually our institutions change too.

Imagine starting with every possible frequency of sound, played altogether. What a cacophony! It's like mixing all the paints together on the page—just a brown mush. But now remove certain frequencies in just the right amounts, and what is left behind is beautiful music.

265

What music are we creating? It's a function of which filters we apply to the wholeness of who we are. But we don't need to create our lives from scratch; who we want to be is already within us. As we face the practical decisions of our lives, we can embody what we value and carve away what we do not. When we face problems, instead of asking, "What do I need to do to solve this?" we can ask, "What do I need to *not* do in order to solve this?" A shift to wholeness helps us let go of our reactionary actions and get into flow.

This applies equally to ourselves as individuals and to our whole community. Rather than a splintered collection of pieces, we come from a single root. We feel isolated from each other and from ourselves only through the choices we have made. But we are one whole. We are similar to each other in our experience of vulnerability and our yearning for satisfaction, and through the ongoing power of our choices, we have the capacity to rediscover our wholeness as a community.

Because the wholeness of the hologram infuses the very space and time you swim in, it must be more than just a great metaphor. The wholeness of space and time is a crucial clue to the wholeness that you are. Seeing the world from the perspective of wholeness is not more correct than seeing the world through its parts. It is a complementary perspective. You are both a whole human being and a collection of parts. When you trim your fingernails, you do not worry about throwing away the cut nails. You do not have a squeamish sense that you have cut off a part of yourself and are now less than you were. You have an identity in total that doesn't exist for the individual parts. Your identity exists *as-a-whole*, composed of the parts but not identical to them.

For this reason, when you make a mistake, you are not the mistake. If you identify with the mistake, you are bound to repeat it. If you can find an "inner room" of compassion and acknowledge the wound that led you astray, you are identifying with the core of who you are, and you may find it easier to change.

Yet the parts obviously have value in making us who we are. When I habitually bite my nails, I have a different sense of myself than when

I have the presence of mind to let my nails grow. When I don't bite my nails, I feel more confident in every part of my life. The things that happen to our parts affect who we are as-a-whole. Our identity is influenced by the mundane choices we make.

Does this apply collectively too? We are one whole planetary organism. We share molecules of air and water, and we share feelings and thoughts in feeling-space. Our collective responses to world events or social memes have a life of their own, transcending any one individual's reaction. We have become a collective organism. How do we navigate these choices? How do we form our personal and collective identity out of the multitude of possibilities available to us in the holographic multiverse? The leap to wholeness is the "yes" that we say to life. It is not a verbal yes but a physical yes. It is the moment when we see what we risk by doing things differently and we take the risk anyway. It is the moment when understanding our filters becomes transmuted into action in the real world. When we confront fear and make courageous choices, we are embracing new aspects of who we are. When I say yes to sharing a piece of art that feels imperfect, or say yes to an opportunity to have fun with people I love even when I am worried about work, or say yes to trying a new line of work, I am saying yes to a part of myself that feels vulnerable and hidden. I am expanding my known self into new, darker areas of my inner world. I am embracing the whole territory of me. This is a leap to wholeness.

The science we discussed is a field called *signal processing*. It is the study of how information is transformed from one expression into another, how some information is removed while other information is preserved. In signal processing, the manipulations that we perform on our information content are called *filters*. Filtering is how we get from white light to the individual colors that make up a rainbow. Filters take us from the possibilities of who we could be to the actuality of who we are. The more inner filters we can remove, the more we can respond from the depths of compassion and creativity that really define us.

Which Filmstrip Are We On?

Nobody escapes filters. Everybody develops armor in their childhood. Remember Robin DiAngelo's insight about people: "All humans have prejudice; we cannot avoid it."[5] We have patterns of reactivity—shame, hurt, anger, self-criticism, resentment—that do not help our cause. I wouldn't say that these are the *reasons* for our problems. We are not to blame for our circumstances. But they get in the way of making things better.

Can you imagine a world that is peaceful and just yet harbors such unexpressed pain under the surface? It seems unlikely to me. If we want that other path, our pain is going to have to heal.

When we decide to acknowledge and heal our inner reactivity, we put ourselves on a trajectory where useful opportunities may be more likely to happen to us. These are called *synchronicities*, and they serve to help us grow into the people we become. They are not positive but self-reinforcing reflections of who we are. The cosmos continually reinforces our belief through the synchronicities that come our way. Where then does change come from? Novelty comes from inside of us. It is only through noticing how we have messed up our lives that something inside of us says, "No! I don't want to do that anymore." The world has no preference. It will keep sending us more of the same thing that we put out. If we are manipulative or abrasive, we will continue to have the same problems show up in our lives. But one day, having seen the pattern enough times, we may say, "No! I don't want to do that anymore." The power for change is that spark within us.

We live not in moments but in timelines. Which filmstrip do we decide to watch? Which dream do we choose to live? The past and future are inextricably linked, and our choices tug on the timelines to influence the quality of the present.

Although we don't get to choose where we begin in life, the circumstances we experience right now are part of a chain of events extending from the past. If we can find a way to fit our daily challenges into a broader, more meaningful context, we can grow more quickly to the next level of challenge.

Yet I wrestle with this concept. Knowing that some people suffer under systems of oppression that they cannot control, I feel unclear about how my own quest for empowerment can be helpful to someone else. Some people are given tremendously difficult circumstances to handle. Their choices may be more constrained than mine. Racism is the institutionalization of prejudice that systematically undermines the efforts of Black, Indigenous, and People of Color (BIPOC) to advance. Our circumstances do not always reflect our choices.

Maybe we can narrow our lens away from the most difficult experiences and focus on the more personal aspects of choice. Maybe we can look at how the impact we have on others reflects our choices. Regardless of our given circumstances, synchronicity shows up in our day-to-day interactions, and our choices are reflected in so many ways. We can see ourselves in our relationships, we can see ourselves in our work, we can see ourselves in our struggles, we can see ourselves in our family. Our circumstances are not a judgment of our choices. They are a way for us to see who we are. Or sometimes they allow us to see who we are not.

In *Living in Flow* I stated, "The inner experience of the organism is the content of what is being communicated, and the literal experiences—the physical circumstances—are the medium of communication."[6] I don't think this is true for each situation happening in the world. We should try to interpret every single thing as meaningful.

The physical world has a tendency to mirror what we feel. All of us have a wound on our soul. We all live within a system of thinking and feeling that is broken and hurts us. Whether or not it hurts us directly, we suffer together from the incomplete communication that perpetuates misunderstanding.

It's not that the outer causes the inner, and it's also not that the inner causes the outer. Rather, they are reflections of one another. When I hear of an innocent person killed or violence in the streets, my soul contains that violence, too. It's not that I am experiencing or feeling the same specific circumstances. It's that I feel the wound of separateness and violence. I feel your wounds and you feel mine because they are part of a public trust, a collective unconscious that can't be separated into individual people. It shows up as racism. It shows up as bigotry. It shows up as political obstruction. It shows up as domestic violence, or verbal abusiveness, or self-inflicted harm. It is the separateness that we all share. This is where humanity is right now.

Wholeness is not about our circumstances. Wholeness is about knowing ourselves. This we have control over. This is the healing and growth that synchronicity can serve. It is a mistake to think that the outside reflects the inside *literally*. It is not the content but the context of our lives that reflects us. We cannot explain every event that happens. We cannot make sense of violence or malice, injustice or hurtfulness. The leap to wholeness is about opening ourselves to life's suffering. We can take inspiration from those who have stood strong in their truth to help us learn and grow. Life does not reward our efforts with a commensurate salary. Our suffering is the gift we offer to the universe in our effort to bring wholeness to ourselves and our communities.

Healing is available to us, and that path lies open to us. Charles Eisenstein implores, "It is not just visionaries who have seen it. You, dear reader, have surely seen it too, bobbing in and out of sight as you struggle to keep your head above the choppy waters of habit and doubt. We are here to remind each other that it is there for the choosing."[7]

My hope is that we can see the dominos that fall on each other, each event leading to the next, and we can see the role that our own passions play in the collapse of those dominos. Whether it is a family argument or a national crisis, we are all contributing to the growth, or decay, of wholeness in our community.

We are here to do the work that is needed in the world at each moment. We have no idea what that work is, and we cannot guess. We can follow our own inner creative impulse and address each situation that arises. If the world is responsive, we can be assured that what comes to us is what needs to be accomplished right now. It's not for us. We are for it.

We live not just in the moment but on a timeline, a wave moving from our past into our future. The cosmos is like a dream. It is a made of ethereal forms, holograms that present themselves in certainty to us at one moment but become uncertain after we look away. Every choice we make ripples throughout the ocean of information underlying this theater of physical forms. Nothing is missed.

We are only here for a while. Our choices matter. Which dream will we choose?

TESTABILITY

The conjecture that the world is like a hologram is testable. If the only things we can call "real" are the fence posts or ladder rungs, then our world cannot be both objective and certain.

For instance, I claim that the hiring decision conveyed in an email, time-stamped at noon, only became part of my fence-post reality at 6 p.m. At this time the decision became a fact for me, and before that the email message lived in uncertainty, along with the author of the email. Did they or did they not send the email?! Their reality must remain uncertain, from my perspective, because the contents of the email they already wrote are, from my point of view, still flexible. We can be certain of our own reality because we ourselves experienced it. Therefore, the certainty we experience must be subjective.

For the holographic model of the cosmos to be consistent, what we call *reality* has to depend on perspective. Physical facts must be, in some sense, subjective. There can be no objective, definite world, only a world experienced from your perspective and required to be consistent with everyone you meet.

In quantum mechanics, there is a testable difference between subjective and objective events. We'll call the subjective description *relative reality*. That the world is described in this relative manner is a proposal put forward originally by Hugh Everett with the Many Worlds hypothesis, then by Eugene Wigner with the Wigner's Friend thought experiment, and more recently by Carlo Rovelli and others with Relational Quantum Mechanics.

Far from being far-fetched, this approach is the Occam's razor—or simplest—way to think about quantum mechanics. The mathematics of quantum mechanics in its purest form points to an ever-expanding tree of possible realities, rather than one actual reality. In the relative reality or "relationality" model, what we call an "actual" reality is only relatively true.

This hypothesis has been tested, and although the tests are abstract, they point in the right direction. I'll list two recent proposals that have been made to test relationality, one of which was experimentally performed and matched expectations.

Imagine I flip a coin. The goal is to see whether *my* measurement of the coin (heads or tails) limits the results you can get when you look a moment later. You might think it would, but in fact the theory implies differently. What we expect in a holographic multiverse is that when I measure the coin, I simply expand my rendered world to include the coin. But for you, arriving a moment later, both the coin and I are unrendered. The coin and I are connected, because I have looked at it. But you don't yet know what I saw. Therefore, for you, both options—heads or tails—are possible. If you look at the coin and see heads, then you know for sure that I will also say I measured heads. We must have consistent results, or we'd say the virtual reality was broken. But there is also the possibility that you see tails, and when you check with me, so have I. My reality has fallen into place after the fact.

Massimiliano Proietti and others tested this using six entangled photons to represent two separate laboratories.[1,2] We'll call them Alice's and Bob's laboratories, and Alice and Bob each have a Friend in the lab. There is a photon each to represent, Alice, her Friend, and the System they are measuring. The same goes for Bob, for a total of six. As a precondition, the two Systems are set up to be entangled. By doing the appropriate interactions and manipulations on the photons, the authors check whether Alice's measurement of her photon preconditioned the results that Bob would get if they measured the same photon afterward.

Proietti found that the results are consistent with "no-collapse" quantum evolution: the System and the Friend remain in a superposition state even with confirmation of Alice's Friend's definite measurement. What does this mean?

Their analysis argues that modern physics relies on three basic assumptions. The first assumption is *locality*, the premise that things affect only things near them. In other words, we can't send a signal instantly to another galaxy. The second assumption is *free choice*, the premise that we can choose our measurements freely. This means nature has some fundamental unpredictability, since sentient beings like you and me make choices that cannot be predicted by the laws of physics. The final assumption is *observer-independence*, which is the notion we are talking about in this appendix. This assumption is that there is a truth or falseness about who has interacted with whom. According to this assumption, if I measure the coin first and I get heads, then that is the truth for everybody.

The results obtained in the experiment imply that at least one of the earlier assumptions must be false. My preference, of course, is that the third, observer-independence, is false. This is what we would expect from a holographic multiverse, or any version of quantum mechanics in which collapse of the wavefunction is a relative or relational process.

Again, the difference is subtle. In a relational universe, if I measure the coin first and I get heads, then from my perspective everyone else will agree with me. But from their perspective, who can say whether I got heads or not? If you come along and measure tails, then I will have measured tails too. This is a different branch of the multiverse, a different branch of the tree of possibilities. In the relational universe, there is no bird's-eye view that judges what really happened. Any possible description is from the view of somebody or something, and each person can only see what they can see. From every perspective, knowledge of the world is incomplete.

A related paper by Daniela Frauchiger and Renato Renner demonstrated a similar conclusion.[3] They propose an experiment that shows that quantum mechanics generates inconsistent experimental predictions if, again, three similar assumptions are all held to be true: universality, consistency, and objectivity. In this proposed experimental setup, the results of a quantum measurement—say, the spin direction of an electron or the polarization direction of a photon—will lead to logical contradictions between measurements in one-twelfth of the runs of the experiment.

The logic in their paper is a bit difficult to wade through, but one conclusion that could resolve the dilemma is again dropping objectivity. A system can have different values for different people, but only in different branches of the multiverse. In a given branch—a given universe— the observers have to agree. Hence consistency—our agreement on things we experience in common—is more important than objectivity. As we've discussed, the holographic model requires quantum mechanics to apply to everything, at all scales, and there are no special rules for large versus small objects. This is what is meant by *universality*.

A summary from these experiments, both reasonably respected within the academic physics community, is that if we accept a relational multiverse and abandon the belief in objectivity, we can experience a shared reality that makes sense and is consistent, where we do really have free choice as we feel we do, we can't instantly communicate across the galaxy, and there aren't different rules for atoms than there are for the people and things made of atoms.

Although it's taken many decades for the relational view to gain traction given the challenges it poses to our concrete sense of reality, over time it is one of only a handful of basic approaches that has survived.

B

REVIEW OF MEANINGFUL
HISTORY SELECTION

Meaningful history selection is a process proposed in my first book, *Living in Flow*, that accounts for experiences of synchronicity. If you have not read the first book or do not recall the basic points, I will summarize them here.

In quantum mechanics, the universe is described by wavefunctions, which allow for several different mutually exclusive descriptions of a physical reality to be combined into one complete description. In fact, each time a wavefunction is measured, it may split into many such versions of reality. Thus, over time the complexity of the cosmos—the number of possible arrangements of things—grows.

The natural way to draw these branches is in the form of a tree. Each branch of the tree represents *all of the universe* branching into different versions of itself. In other words, it is *a decision tree through possible experiences*. On each branch is a complete version of everything in the world, just organized differently from the versions of our world on the other branches. This is often called the Many Worlds interpretation of quantum mechanics.

So far, although not everyone agrees that this is the best interpretation of quantum mechanics, there is general agreement that the math is correct.

My own interest is in studying how these branches can be flexible.

Let's use one tree to represent you and a separate tree to represent me. Then we can investigate how your branches meet up with mine when we interact. Our trees' branches are intersecting. How do our worlds overlap? Well, the top of the tree represents later times and the bottom of the tree represents earlier times. Before I get to the height where my tree branch meets yours, do I know anything about you? No. It is only the point where our branches touch that I can look down your branch and say, "Ah! I know all about you now!" Before that, I had to claim ignorance.

I call this *retroactive event determination*. All the things you've done become clear to me only after we meet.

Now, I propose that the experience we are having—the context—is more fundamental than the physical world—the content. In this case it matters what I am anticipating when I take action. My actions reflect the experiences I want to have, and since the world is a physical expression of the experience, then my anticipation of an experience has an effect on the probabilities of physical circumstances. In other words, my *anticipated qualitative experience* influences the likelihood of which branch of yours I meet when I climb the tree.

While trying to work out these ideas, I was talking with my wife, Dana, who remarked that the anticipation of certain experiences seemed to create a "weight factor." Through this process, one naturally biases the probability of some branches more than others. The idea of apples on a tree came to us, where some branches have heavy apples, some have light apples, and some have none at all, as in Figure B.1. From my subjective perspective, as I take action while anticipating certain outcomes, I make it more likely that meaningful circumstances will unfold that lead to the experience I anticipate. This is meaningful history selection.

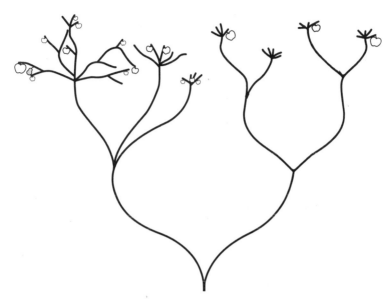

FIGURE B.1. In this theory, a branching tree of possible life events describes your experience. You can always think of yourself as looking up at a branching tree of options above you. The apples represent situations you anticipate through the choices you make. The apples make certain branches more likely. A synchronicity is any event that becomes more likely because it leads to an experience (apple) you are seeking to have.

REVIEW OF THEORIES OF THE HOLOGRAPHIC MULTIVERSE

Research into the possibility of a holographic multiverse has a long history. Physicist Hugh Everett III introduced the idea of a multiverse with the Many Worlds interpretation of quantum mechanics in the 1950s. While largely ignored at first, this theory has gained favor over time. Although the ontological idea of many *actual versions of you and me* is not widely adopted, from a mathematical perspective it forms a pillar of our modern understanding of quantum mechanics.

Whether one should think of Everett's interpretation as many actual universes or just as a handy calculational tool has been explored more recently by Anthony Aguirre. His and others' work examines, for instance, whether one could test for the existence of other universes, whether two such universes could collide, and other similar questions that put the idea on a firm investigative footing.

In the 1990s Leonard Susskind and Gerard 't Hooft introduced the idea of "the universe as a hologram." Their concept is based on a result from black holes that indicates that all the three-dimensional information of objects that fall into a black hole exists in two dimensions on its surface; they liken this to a hologram, which shows a three-dimensional object within a two-dimensional piece of film.

Juan Maldacena furthered these ideas with the AdS/CFT correspondence; his original paper on the subject is the most frequently referenced paper in the history of physics. This paper demonstrated an equivalence between string theory in five dimensions (connected to theories of quantum gravity) and quantum field theory in four dimensions. Again, as in a hologram, we have a higher-dimensional space encoded in a lower dimension.

Finally, Stephen Hawking and a collaborator more recently introduced a model for a holographic multiverse. A common motivation in the quest to understand the multiverse is to explain why there are such special values for the fundamental constants in our universe, such as the speed of light. Could there be other universes with different values for these constants? Hawking's approach incorporates the holographic universe idea from string theory into traditional theories of the Big Bang. The theory predicts that the Big Bang could give rise to many pocket universes, each with its own properties.

The approach I take in my research and in this book may be compatible with these other approaches, but where it aligns and where it diverges is an active subject of research.

NOTES

Preface

1 Winnicott, "Ego Distortion," 140–52.
2 Satchidananda, "Balance with Maya and Adversity."

Chapter 1

1 Sagan, *Contact*.
2 Jung, *Synchronicity*.

Chapter 2

1 Capra and Luisi, *Systems View of Life*, 133.
2 Capra and Luisi, *Systems View of Life*.
3 Capra and Luisi, *Systems View of Life*, 133.
4 Capra and Luisi, *Systems View of Life*, 134.
5 Capra and Luisi, *Systems View of Life*, 129.
6 Capra and Luisi, *Systems View of Life*, 132.
7 Capra and Luisi, *Systems View of Life*, 136.
8 Maturana and Varela, *Tree of Knowledge*.
9 Eisenstein, *Climate*, 8.
10 Eisenstein, *Climate*, 42.
11 Nussenzveig, "Theory of the Rainbow."
12 Mills, "Physics of Rainbows."

Chapter 3

1 Nelson, "Macroscopic Quantum Superposition States."
2 Nelson, "Relativistic Roots."
3 Nelson-Isaacs, "Guiding Quantum Histories."

4 Nelson-Isaacs, "Holographic Framework for Wavefunction Propagation."
5 Napolitano and Sakurai, *Modern Quantum Mechanics*, 55.
6 Goodman, *Introduction to Fourier Optics*.
7 Lewis, "Light Waves and Light Corpuscles."
8 Lewis, "Nature of Light."
9 Jaworski, *Synchronicity*, 182.

Chapter 4

1 Valtorta et al., "Loneliness and Social Isolation Coronary."
2 Adamson and Pennycook, "Trapped in a Bubble."
3 Holt-Lunstad et al., "Loneliness and Social Isolation Mortality."
4 Clarke et al., *Use of Yoga, Meditation, and Chiropractors.*
5 Banyaca, Sr., "Prophecy."
6 Garavito and Thanki, "Stop Asking People of Color."
7 Bobby, "Reclaim the Power."
8 Bobby, "Reclaim the Power."
9 Jung, *Synchronicity*, 35.

Chapter 5

1 Huggett, "Zeno's Paradoxes."
2 Csikszentmihalyi, *Flow*.

Chapter 6

1 Krishnananda, *Concept of Maya*.
2 Krishnananda, *Study and Practice of Yoga*.
3 Krishnananda, *Study and Practice of Yoga*.
4 Yogananda, *God Talks with Arjuna*.

Chapter 7

1 Singer, *Untethered Soul*, 61.
2 Marshall, "Bringing Ourselves to Work."
3 Singer, *Untethered Soul*, 61.
4 Aristotle, *Nicomachean Ethics* 1109a.
5 Haines, *Politics of Trauma*.
6 Singer, *Untethered Soul*, 145.

7 Singer, *Untethered Soul*, 151.

8 Roosevelt, *Inaugural Address*.

Chapter 8

1 Chodron, *When Things Fall Apart*, 66.

2 Haines, *Politics of Trauma*.

3 Chodron, *When Things Fall Apart*, 12.

4 Satchidananda, *Yoga Sutras of Patanjali*.

5 Chodron, *When Things Fall Apart*, 78.

6 Chodron, *When Things Fall Apart*, 14.

7 Wigner, "Unreasonable Effectiveness of Mathematics, p. 14."

8 Chodron, *When Things Fall Apart*, 15.

9 Chodron, *When Things Fall Apart*, 16.

Chapter 9

1 Brown, *Braving the Wilderness*, 154.

2 Brown, *Daring Greatly*, 129.

Chapter 10

1 Eisenstein, *Climate*, 173.

Chapter 11

1 Chodron, *When Things Fall Apart*, 69.

2 Chodron, *When Things Fall Apart*, 69.

3 Herbert, *Lady Gaga: Behind the Fame*.

4 *Monster Ball Tour*.

5 Eisenstein, *Climate*, 42.

Chapter 12

1 DiAngelo, *White Fragility*, 19.

2 DiAngelo, *White Fragility*, 20.

3 DiAngelo, *White Fragility*, 21.

4 Greenberg, "10 Examples."

5 DiAangelo, *White Fragility*, 25.

6 Klein, *Our Need for Others*.

7 Gabel, *Desire for Mutual Recognition*, 131.

8 Gabel, *Desire for Mutual Recognition*, 132.

9 Gabel, *Desire for Mutual Recognition*, 133.

Chapter 13

1 Feynman, "Pleasure of Finding Things Out."

2 Goodman, *Introduction to Fourier Optics*.

3 "Personal Pinhole Theater."

4 Pribram, "Holographic Hypothesis of Brain Function."

5 Pribram, "Quantum Holography."

6 Pribram, "Brain and Perception."

7 Pribram, "What Is All the Fuss About?"

8 Pribram, "Holographic Hypothesis of Brain Function."

9 Campbell, Nachmias, and Jukes, "Spatial-Frequency Discrimination in Human Vision."

10 Pribram, "Holographic Hypothesis of Brain Function."

11 Pribram, "What Is All the Fuss About?"

12 Cooper, Kensinger, and Ritchey, "Memories Fade."

13 Brown, "Some Tests of the Decay Theory."

14 Peterson and Peterson, "Short-Term Retention."

15 Kurzweil, *How to Create a Mind*, 38.

16 Pribram, "Holographic Hypothesis of Brain Function."

Chapter 14

1 Tyson, "Afterlife."

2 Nelson-Isaacs, "Holographic Framework."

3 Bohm, *Wholeness and the Implicate Order*, 190.

4 Bohm, *Wholeness and the Implicate Order*, 190.

5 Bohm, *Wholeness and the Implicate Order*, 76.

6 Bohm, *Wholeness and the Implicate Order*, 188.

7 Bohm, *Wholeness and the Implicate Order*, 189.

8 Bostrom, "Are You Living in a Computer Simulation?," 1.

9 Bostrom, "Are You Living in a Computer Simulation?," 11.

10 Einstein, Podolski, and Rosen, *Quantum Mechanical Description of Physical Reality*, 777.

11 Thorne, "Biographical Memoirs: John A. Wheeler."
12 Thorne, "Biographical Memoirs: John A. Wheeler."

Chapter 15

1 Rumi, *Essential Rumi.*
2 Scharmer, *Essentials of Theory U,* xii.
3 Scharmer, *Essentials of Theory U,* 10.
4 Jobs, "Stanford Commencement Address."
5 DiAangelo, *White Fragility,* 20.
6 Nelson-Isaacs, *Living in Flow,* 40.
7 Eisenstein, *Climate,* 174.

Appendix A

1 Proietti et al., "Experimental Rejection of Observer-Independence."
2 Bong et al., "Wigner's Friend's Experience."
3 Frauchiger and Renner, "Quantum Theory Cannot Consistently Describe."

BIBLIOGRAPHY

Adamson, Mike, and Richard Pennycook. *Trapped in a Bubble: An Investigation into Triggers for Loneliness in the UK.* British Red Cross. 2016. www.red-cross.org.uk/-/media/documents/about-us/research-publications/health-and-social-care/co-op-trapped-in-a-bubble-report.pdf.

Banyaca Sr., Thomas, Speaker of the Wolf, Fox and Coyote Clan Elder of the Hopi Nation. "Prophecy: We Are the Ones We've Been Waiting For." 2011. www.awakin.org/read/view.php?tid=702.

Bekenstein, Jacob D. "Black Holes and Entropy." *Physical Review D* 7 (1973): 2333–46. https://doi.org/10.1103/PhysRevD.7.2333.

Bobby. "Open Letter from the Wretched of the Earth Bloc to the Organisers of the People's Climate March of Justice and Jobs." Reclaim the Power. January 17, 2016. https://reclaimthepower.org.uk/news/open-letter-from-wretched-of-the-earth-bloc-to-organisers-of-peoples-climate-march/.

Bohm, David. *Wholeness and the Implicate Order.* New York: Routledge, 1980.

Bong, Kok-Wei, Aníbal Utreras-Alarcón, Farzad Ghafari, Yeong-Cherng Liang, Nora Tischler, Eric G. Cavalcanti, Geoff J. Pryde, and Howard M. Wiseman. "Testing the Reality of Wigner's Friend's Observations." arXiv. 2019. https://arxiv.org/abs/1907.05607.

Bostrom, Nick. "Are You Living in a Computer Simulation?" *Philosophical Quarterly* 53, no. 211 (2003): 243–55. https://doi.org/10.1111/1467-9213.00309.

Brown, Brené. *Daring Greatly: How the Courage to Be Vulnerable Transforms the Way We Live, Love, Parent, and Lead.* New York: Gotham Books, 2012.

Brown, John. "Some Tests of the Decay Theory of Immediate Memory." *Quarterly Journal of Experimental Psychology* 10, no. 1 (1958): 12–21. https://doi.org/10.1080/17470215808416249.

Campbell, Fergus W., Jacob Nachmias, and John Jukes. "Spatial-Frequency Discrimination in Human Vision." *Journal of the Optical Society of America* 60, no. 4 (1970): 555–59. https://doi.org/10.1364/JOSA.60.000555.

Capra, Fritjof, and Pier Luigi Luisi. *The Systems View of Life: A Unifying Vision.* Cambridge, UK: Cambridge University Press, 2014.

Chodron, Pema. *When Things Fall Apart.* Boston: Shambhala, 1997.

Clarke, Tainya C., Patricia M. Barnes, Lindsey I. Black, Barbara J. Stussman, and Richard L. Nahin. *Use of Yoga, Meditation, and Chiropractors among U.S. Adults Aged 18 and Over.* US Department of Health and Human Services, Centers for Disease Control and Prevention, National Center for Health Statistics. 2018. www.cdc.gov/nchs/data/databriefs/db325-h.pdf.

Contact. Directed by Robert Zemeckis. Performed by Ellie Arroway. Warner Bros., 1997.

Cooper, Rose A., Elizabeth A. Kensinger, and Maureen Ritchey. "Memories Fade: The Relationship Between Memory Vividness and Remembered Visual Salience." *Psychological Science* 30, no. 5 (2019): 657–68. https://doi.org/10.1177/0956797619836093.

Csikszentmihalyi, Mihaly. *Flow: The Psychology of Optimal Experience.* New York: Harper & Row, 1990.

DiAngelo, Robin. *White Fragility.* Boston: Beacon Press, 2018.

Einstein, Albert, Boris Podolski, and Nathan Rosen. "Can Quantum-Mechanical Description of Physical Reality Be Considered Complete?" *Physical Review* 47, no. 10 (1935): 777–80. https://doi.org/10.1103/PhysRev.47.777.

Eisenstein, Charles. *Climate: A New Story.* Berkeley, CA: North Atlantic Books, 2018.

Feynman, Richard. "The Pleasure of Finding Things Out." British Broadcasting Corporation. 1981. www.bbc.co.uk/programmes/p018dvyg/clips.

Frauchiger, Daniela, and Renato Renner. "Quantum Theory Cannot Consistently Describe the Use of Itself." *Nature Communications* 9, article no. 3711 (2018). https://doi.org/10.1038/s41467-018-05739-8.

Gabel, Peter. *The Desire for Mutual Recognition.* New York: Routledge, 2018.

Garavito, Tatiana, and Nathan Thanki. "Stop Asking People of Color to Get Arrested to Protest Climate Change." Vice. September 18, 2019. www.vice .com/en_us/article/mbm3q4/extinction-rebellion-xr-is-shaped-by-middle -class-white-people-it-does-not-serve-people-of-color.

Glick-Smith, Judith. *Flow-Based Leadership: What the Best Firefighters Can Teach You about Leadership and Making Hard Decisions.* Basking Ridge, NJ: Technics, 2016.

Goodman, J. W. *Introduction to Fourier Optics.* New York: W. H. Freeman, 2004.

Greenberg, Jon. "10 Examples That Prove White Privilege Exists in Every Aspect Imaginable." *Yes,* July 24, 2017. www.yesmagazine.org/social-justice /2017/07/24/10-examples-that-prove-white-privilege-exists-in-every -aspect-imaginable/.

Haines, Staci K. *The Politics of Trauma: Somatics, Healing, and Social Justice*. Berkeley: North Atlantic Books. 2019.

Herbert, Emily. *Lady Gaga: Behind the Fame*. New York: Overlook, 2010.

Holt-Lunstad, Julianne, Timothy B. Smith, Mark Baker, Tyler Harris, and David Stephenson. "Loneliness and Social Isolation as Risk Factors for Mortality: A Meta-Analytic Review." *Perspectives on Psychological Science* 10, no. 2 (2015): 227–37. https://doi.org/10.1177/1745691614568352.

Huggett, Nick. "Zeno's Paradoxes." *Stanford Encyclopedia of Philosophy*. April 30, 2002; revised June 11, 2018. https://plato.stanford.edu/entries/paradox-zeno/.

Jaworski, Joseph. *Synchronicity: The Inner Path of Leadership*. Oakland, CA: Berrett-Koehler, 2011.

Jobs, Steve. "Stanford Commencement Address." Filmed June 12, 2005, at Stanford University, Stanford, CA. www.youtube.com/watch?v=UF8uR6Z6KLc.

Jung, Carl Gustav. *Synchronicity: An Acausal Connecting Principle*. Princeton, NJ: Princeton University Press, 1952.

Klein, Josephine. *Our Need for Others*. London: Tavistock, 1994.

Krishnananda, Swami. *The Concept of Maya According to Saiva Siddhanta*. n.d. Accessed July 9, 2020. www.swami-krishnananda.org/disc/disc_152.html.

Krishnananda, Swami. "Chapter 61: How the Law of Karma Operates." *The Study and Practice of Yoga: An Exposition of the Yoga Sutras of Patanjali*. n.d. Accessed July 9, 2020. www.swami-krishnananda.org/patanjali/raja_61.html.

Kurzweil, Ray. *How to Create a Mind: The Secret of Human Thought Revealed*. New York: Penguin, 2012.

Lady Gaga Presents the Monster Ball Tour: At Madison Square Garden. Directed by Laurieann Gibson. Performed by Lady Gaga. Mermaid Films/HBO, 2011.

Lewis, Gilbert N. "Light Waves and Light Corpuscles." *Nature* 117 (1926): 236–38. https://doi.org/10.1038/117236a0.

Lewis, Gilbert N. "The Nature of Light." *Proceedings of the National Academy of Sciences of the United States of America* 12, no. 1 (1926): 22–29. https://doi.org/10.1073/pnas.12.1.22.

Maldacena, Juan. "The Large N Limit of Superconformal Field Theories and Supergravity." *Advances in Theoretical and Mathematical Physics* 2, no. 2 (1997): 231–52. https://doi.org/10.1023/A:1026654312961.

Marshall, Bowen T. "Bringing Ourselves to Work: A Narrative Inquiry of LGBTQ Professionals." PhD diss., Ohio State University, 2017. https://etd.ohiolink.edu/.

Maturana, Humberto, and Francisco Varela. *The Tree of Knowledge*. Boston: Shambhala, 1998.

Mills, Dick. "Exploration into the Physics of Rainbows." Physics Forum Insights. January 5, 2016. www.physicsforums.com/insights/rainbows-not-vampires/.

Napolitano, J. J., and J. J. Sakurai. *Modern Quantum Mechanics*. New York: Pearson, 2011.

Nelson, Sky. "Retroactive Event Determination and Its Relativistic Roots." In *Quantum Retrocausation-Theory and Experiment*, edited by Daniel P. Sheehan, 45–74. San Diego: AIP Conference Proceedings, 2011. https://doi.org/10.1063/1.3663717.

Nelson, Sky. "Retroactive Event Determination and the Interpretation of Macroscopic Quantum Superposition States in Consistent Histories and Relational Quantum Mechanics." *Journal for Scientific Exploration* 25, no 2 (2011): 273–304.

Nelson-Isaacs, Sky. "A Holographic Framework for Wavefunction Propagation." 2020. Unpublished.

Nelson-Isaacs, Sky. "Guiding Quantum Histories with Intermediate Decomposition of the Identity." *AIP Conference Proceedings* 1841 (2016). https://doi.org/10.1063/1.4982770.

Nelson-Isaacs, Sky. *Living in Flow: The Science of Synchronicity and How Your Choices Shape Your World*. Berkeley, CA: North Atlantic Books, 2019.

Nussenzveig, H. Moyses. "The Theory of the Rainbow." *Scientific American*, April 1977, 116–28.

"Personal Pinhole Theater." Exploratorium. n.d. Accessed May 25, 2020. www.exploratorium.edu/snacks/personal-pinhole-theater.

Peterson, L., and M. J. Peterson. "Short-Term Retention of Individual Verbal Items." *Journal of Experimental Psychology: General* 58, no. 3 (1959): 193–98. https://doi.org/10.1037/h0049234.

Pribram, Karl. *Brain and Perception: Holonomy and Structure in Figural Processing*. Mahwah, NJ: Lawrence Erlbaum, 1991.

Pribram, Karl. "The Holographic Hypothesis of Brain Function: A Meeting of Minds." 1984. http://karlpribram.com/wp-content/uploads/pdf/theory/T-148.pdf.

Pribram, Karl. "Quantum Holography: Is It Relevant to Brain Function?" *Information Sciences* 115 (1999): 97–102. https://doi.org/10.1016/S0020-0255(98)10082-8.

Pribram, Karl. "What Is All the Fuss About?" In *The Holographic Paradigm and Other Paradoxes*, edited by Ken Wilber. Boulder, CO: Shambhala, 1982.

Proietti, Massimiliano, Alexander Pickston, Francesco Graffitti, Peter Barrow, Dmytro Kundys, Cyril Branciard, Martin Ringbauer, and Alessandro Fedrizzi. "Experimental Rejection of Observer-Independence in the Quantum World." *Science Advances* 5, no. 9 (2019): eaaw9832. https://doi.org/10.1126/sciadv .aaw9832.

Roosevelt, Franklin Delano. *Inaugural Address.* National Archives. January 1933. www.archives.gov/education/lessons/fdr-inaugural.

Rumi, Jellaludin. *The Essential Rumi.* Edited and translated by Coleman Barks. New York: HarperCollins, 1995.

Satchidananda, Sri Swami. "How to Find Balance with Maya and Adversity." Interview by Integral Yoga. n.d. Accessed July 9, 2020. www.youtube.com/ watch?v=9mKL6dJUNgQ.

Satchidananda, Sri Swami. *The Yoga Sutras of Patanjali.* Buckingham, VA: Satchidananda Ashram-Yogaville, 1978, repr. 2012.

Scharmer, Otto. *The Essentials of Theory U: Core Principles and Applications.* Oakland, CA: Berrett-Koehler, 2018.

Singer, Michael A. *The Untethered Soul.* Oakland, CA: New Harbinger, 2007.

Susskind, Leonard. "The World as a Hologram." *Journal of Mathematical Physics* 36, no. 11 (1994): 6377–96. https://doi.org/10.1063/1.531249.

Talbot, Michael. *The Holographic Universe: The Revolutionary Theory of Reality.* New York: HarperCollins, 1991.

't Hooft, Gerard. "Dimensional Reduction in Quantum Gravity." *AIP Conference Proceedings C* 116 (1993, rev. 2009): 284–96. https://arxiv.org/abs /gr-qc/9310026.

Thorne, Kip. "Biographical Memoirs: John A. Wheeler." NAS Online. 2008. www.nasonline.org/publications/biographical-memoirs/memoir-pdfs /wheeler_john.pdf.

Tyson, Neil deGrasse. "Neil deGrasse Tyson on the Afterlife, Origins of the Earth and Extreme Weather." Interview by Larry King. November 9, 2017. www.youtube.com/watch?v=4x2ZrklQQYU.

Valtorta, Nicole K., Mona Kanaan, Simon Gilbody, Sara Ronzi, and Barbara Hanratty. "Loneliness and Social Isolation as Risk Factors for Coronary Heart Disease and Stroke: Systematic Review and Meta-Analysis of Longitudinal Observational Studies." *Heart* 102, no. 13 (2016): 1009–16. https:// doi.org/10.1136/heartjnl-2015-308790.

Wigner, Eugene. "The Unreasonable Effectiveness of Mathematics in the Natural Sciences." *Communications on Pure and Applied Mathematics* 13 (1960): 1–14. https://doi.org/10.1002/cpa.3160130102.

Winnicott, Donald. "Ego Distortion in Terms of True and False Self." In *The Maturational Process and the Facilitating Environment: Studies in the Theory of Emotional Development*, 140–52. New York: International Universities Press, 1965.

Yogananda, Paramahansa. *God Talks with Arjuna: The Bhagavad Gita.* Los Angeles: Self-Realization Fellowship, 1995.

INDEX

Italic page numbers indicate illustrations